ELECTRODEPOSITION OF METALS IN ULTRASONIC FIELDS

ELEKTROOSAZHDENIE METALLOV V UL'TRAZVUKOVOM POLE

ЭЛЕКТРООСАЖДЕНИЕ МЕТАЛЛОВ В УЛЬТРАЗВУКОВОМ ПОЛЕ

Sergei Mikhailovich Kochergin
Galina Yakovlevna Vyaseleva

Laboratory of Physical Chemistry
S. M. Kirov Chemical Engineering Institute
Kazan, USSR

Translated from Russian

ELECTRODEPOSITION
OF METALS
IN ULTRASONIC FIELDS

$\frac{c}{b}$ Springer Science+Business Media, LLC 1966

ISBN 978-1-4899-4760-4   ISBN 978-1-4899-4758-1 (eBook)
DOI 10.1007/978-1-4899-4758-1

The original Russian text was published by Vysshaya Shkola Press in
Moscow in 1964.

Library of Congress Catalog Card Number 65-26633

FOREWORD

At the present time ultrasound is being utilized on an ever-increasing scale for the deposition of metals. The effect of ultrasound vibrations on the quality of electrolytic deposits is being studied in great detail. Most authors have concluded that electrodeposition of metals by ultrasound vibration can be accelerated considerably by using high current densities. Under these conditions the metal yield increases and the physicomechanical properties of the deposits are improved; the hardness and elasticity increase and the porosity is diminished.

Ultrasonic vibration makes the composition of the electrolytic bath more uniform (parts with complex shapes are coated much more uniformly) and the functioning of the anodes becomes more stable. In a number of processes control of the electrolyte composition is simplified. In the Soviet Union and abroad many laboratory results have been tested industrially. Ultrasonic equipment for degreasing machine parts prior to plating has been installed in assembly lines.

Electrolytic baths with an ultrasonic vibrator mounted in them are being manufactured at the present time. All the work which has been done up to now has dealt with small machine parts. The volumes of the present-day baths in which ultrasonic vibration is used do not exceed 50-100 liters. Plating of larger machine parts may lead to difficulties in the ultrasonic technique and the process of electrodeposition.

The application of ultrasonics promotes progress in the electroplating industry, but it requires the study of the conditions of production and the quality of plating. Up to the present time, the use of ultrasound has been most effective in treating small parts with intricate shapes. It is particularly useful in producing high-quality platings on critical parts. The use of ultrasound in electroplating facilities must be justified economically. The expensive equipment and the additional consumption of electrical power must be compensated by increasing the rate of the process and the quality of the plate.

The material covered in this book refers to the theory and practice of electrodeposition of metals in an ultrasonic field. The sources of this material are the results of original investigations published during the last 15 years and the experimental results obtained since 1953 in the Laboratory of Physical Chemistry of the S. M. Kazan Chemicotechnological Institute.

PUBLISHER'S NOTE

The following Soviet journals cited in this book are available in cover-to-cover translation:

Russian Title	English Title	Publisher
Zhurnal fizicheskoi khimii	Russian Journal of Physical Chemistry	The Chemical Society (London)
Zhurnal prikladnoi khimii	Journal of Applied Chemistry of the USSR	Consultants Bureau

CONTENTS

PHYSICAL BASIS OF THE EFFECT OF ULTRASOUND
ON ELECTRODE PROCESSES

The physical basis of the effect of sonic and ultrasonic vibrations has been treated in detail in many texts and monographs [1-9]. We shall deal here only with the main physical parameters of elastic vibrations in the sonic and ultrasonic range as they affect electrode processes.

The nature of ultrasonic vibrations is the same as that of sonic vibrations. However, because of the high frequencies (and consequently small wavelengths) involved, they have a number of singularities. The propagation of sound waves leads to alternating compression and rarefaction of the medium, the amplitudes of compression and rarefaction being equal. This alternation corresponds to the vibrational frequency of the sound wave. The frequency of the sound wave is measured by the number of vibrations per second. One vibration per second is called a hertz [29]. The frequency of ultrasonic vibrations is between $20 \cdot 10^3$ and 10^9 Hz [29]. The frequency of vibration is related to the wavelength by the following expression:

$$\lambda = \frac{a}{f},$$

where λ is the wavelength, a is the velocity of sound, and f is the frequency of vibration.

The amount of acoustic energy passing per unit time through an area of one square meter perpendicular to the direction of propagation of the sound waves is called the sound intensity.

For a plane wave the sound intensity is determined by the equation

$$I = \omega a = \frac{1}{2} \rho a f A^2 = \frac{1}{2} \frac{\rho a}{f^2} B^2 = \frac{1}{2} \rho a U^2 = \frac{p_0 U}{2},$$

where ω is the average energy density of ultrasonic vibrations, a is the velocity of sound (Table 1) [4], ρ is the density of the medium, U is the particle velocity amplitude of the medium, p_0 is the amplitude of the variable sound pressure, f is the vibrational frequency, A is the vibrational amplitude, and B is the amplitude of acceleration.

TABLE 1. Rate of Propagation of Sound Waves a, Density ρ, and Wave Resistance ρa in Solids

Substance	$\rho \cdot 10^{-3}$, kg/m^3	a, m/sec	$\rho a \cdot 10^3$, kg/m^{2}sec
Aluminum	2.7	6260	169
Bakelite	1.4	2590	36
Brass	8.1	4430	361
Quartz	2.6	5570	145
Magnesium	1.7	2330	93
Copper	8.9	4700	418
Nickel	8.8	4630	495
Polystyrene	1.06	2350	93
Lead	11.4	2160	246
Steel	7.8	5810	476
Glass (crown)	2.6	4660	138

TABLE 2. Values of the Parameters of Ultrasonic Vibrations in Water

f, kc	λ, cm	I, W/m$^2 \cdot 10^{-4}$	P, N/m$^2 \cdot 10^{-4}$
20	7.4	0.4	13.7
20	7.4	2	24.5
50	2.96	0.3	9.81
100	1.48	2	24.5
300	0.49	10	52.9

The sound intensity is expressed in W/m^2. The sound intensity is often characterized by the sound intensity level, measured in decibels.

The vibration of particles of the medium during the propagation of the sound wave creates a variable sound pressure p in the medium which changes periodically with the acoustic frequency:

$$p = p_0 \sin f t,$$

and, therefore,

$$p_0 = f a \rho A, \quad \frac{N}{m^2}.$$

Sometimes the radiation pressure has an important effect. If a plane sound wave of intensity I is incident on an obstacle which completely reflects the sound wave, the wave exerts on the obstacle a pressure

$$p_S = \frac{2I}{a}.$$

If the obstacle partially absorbs sound, then the radiation pressure will be cut in half.

Ultrasonic vibrations generated by a vibrator induce in the adjacent particles of the medium a vibratory motion with respect to their equilibrium position and also a continuous displacement, i.e., a constant flow which is called acoustic streaming. The effect of acoustic streaming is manifest in a strong flow leading to intense mixing of the medium. The acoustic streaming is a hydrodynamic effect due to the viscosity of the medium in which the sound propagates. Sound pressure and acoustic streaming play important roles in creating the fountain effect in liquids. The height of the fountain increases with increasing ultrasonic intensity. The fountain effect leads to the formation of very small liquid droplets, some of which remain in suspension.

The values of these parameters for ultrasonic vibrations in water are given in Table 2.

Intense ultrasonic vibrations propagating through a liquid lead to the formation of ultrasonic cavitation, which is closely related to the sound pressure. Cavitation consists of the formation of discontinuities in the liquids or voids filled with liquid vapor or gas (dissolved in the liquid). The voids implode, thereby creating local shock pressures, which may reach hundreds of atmospheres. The implosion of the cavitational voids produces a local increase in the temperature and electrical discharge. The ultrasonic intensity necessary to create cavitations depends on the vibrational frequency and the nature of the liquid. Cavitation occurs at low and high ultrasonic frequencies. At low frequencies strong cavitation occurs at lower sound intensities. The formation of cavitations at high frequencies requires higher sound intensities.

The effect of cavitation on metals leads to disintegration of its surface, i.e., to dispersion of the surface metal.

Sound waves are absorbed in the media in which they propagate. The absorption of sound waves is due to the viscosity, thermal conductivity, and molecular absorption of the medium. The energy of the sound waves is transformed into thermal energy when the sound waves are absorbed in the medium.

When the intensity of the vibrations is high, this transformation results in a considerable increase in the temperature of the medium.

Ultrasonic waves are diffracted, interfere, are refracted, and undergo internal reflection according to the general laws of wave motion. The effect of sonic and ultrasonic frequencies on substances and the different processes and reactions induced in the substances are determined by the vibrational frequency and the intensity and also by the phenomena of cavitation, sound pressure, and acoustic streaming.

A considerable number of investigations has been published during recent years with regard to the effect of ultrasonic fields on different electrochemical phenomena and processes. In particular, the effect of ultrasound on the electrode potential was investigated in [10-14, 16]. In [15] it was shown that the electrical conductivity and acidity of electrolytes are affected by ultrasound; it was also shown that the Debye effect occurs in electrolytes subjected to an ultrasonic field. A number of investigations have dealt with the effect of ultrasonic fields on the evolution of hydrogen during electrolysis and on its overvoltage. The effect of ultrasonic field on electrolytic separation of deuterium was also studied, and it was found that the deuterium yield increases by 26% under the influence of an ultrasonic field.

Some authors explain the effect of the ultrasonic field on the electrode processes by adsorption. The degree of desorption induced by an ultrasonic field was determined by using radioactive isotopes.

Thus, D. Novchik and B. Pavlevich found that a copper plate coated with the P^{32} isotope becomes completely clean after 5 min in an ultrasonic field. The same authors also noted an increase in the energy of P^{32} ions adsorbed on the copper plate.

Some investigations have concerned the effect of an ultrasonic field on corrosion and anodic oxidation of metals, aluminum in particular. In [12, 17, 19] the authors studied the effect of an ultrasonic field on the passivation and depassivation of metals. Ultrasonic vibrations have a positive effect on chemically deposited metal coatings.

Selenium was deposited on Cu, Cd, and Ni from sulfate solutions by the use of ultrasonic vibrations. The effect of ultrasonic vibrations consists in destruction of passive films on the metal, and the rate of deposition increases because ordinary diffusion is replaced with forced circulation.

Most of the investigations have concerned the effect of ultrasonic fields on the electrodeposition of metals and on oxidation-reduction reactions [10-12, 16-20, 50].

It is well known that the kinetics of electrode processes is determined by the ratio between the concentration (diffusion), chemical, and ohmic polarization. Ultrasonic vibration affects these three polarizations differently. The effect of ultrasonic vibration on electrode processes is determined by the fact that a strong ultrasonic field can change the state of the ionic atmospheres and dehydration energy of molecules, and can excite the electrons in the shells of the reacting particles.

J. Guitton [12] emphasizes the important effect of acoustic vibrations on electrode processes. This author assumes that the differences in the results on the effect of ultrasonic field on electrodeposition of metals are due to experimental conditions in which vibrations with different acoustic intensities were used.

Local overheating, mechanical stresses, electric charges, and cavitational-resonance phenomena affect the electrochemical state of surfaces and the orientation of reacting particles in the surface layer. The sound pressure and acoustic streaming can render the concentration more uniform, remove the gaseous phase and the products of electrochemical reaction from the surfaces of electrodes (electrodeposition and electrooxidation) Also, the sound pressure can change the direction of motion of metal ions toward the electrode surface. As a result, the motion of electrically reduced ions to the cathode can follow not only the lines of force but also a direction at a certain angle determined by the resultant of the electric and acoustic fields.

The frequency of ultrasonic vibrations also plays an important role in electrodeposition. The effect of frequencies within the range of 20 kc to 2-3 Mc on electrode processes has also been studied. The results differ. These experiments showed that short ultrasonic waves leave traces of the wave nodes and antinodes on the the surface of the cathode. The precipitation of the metal is irregular, particularly in the antinodal part of the wave.

Cavitation decreases the concentration and chemical polarizations. Violent mixing resulting from cavitation decreases the limits imposed on diffusion in the bulk of the electrolyte and also at the electrode-electrolyte boundary. The role of cavitation in the removal of the gaseous phase is very important, particularly when the products of electrolysis are gases. The gas bubbles dissolved in the electrolyte are removed to the surface. Mixing is induced by the ascending gas bubbles and by the shock waves.

The effect of cavitation on chemical polarization was not studied in detail. However, it was noted that a strong ultrasonic field affects the behavior of the ions and thus determines the possible mode of variation of the chemical polarization.

The great destructive force of the shock wave and resulting heterogeneities and discreteness of the medium impair the electrocrystallization process, which appears as a change in the ohmic polarization. Cavitation induces dispersion of both the nuclei formed and the growing crystals. Ultrasonic vibrations change the physicomechanical properties of electrolytic deposits and their hardness, elasticity, and porosity by changing the grain size, hydrogen concentration, and internal stresses, and this also proves the effect of ultrasonic vibration on phase polarization. Therefore, ultrasonic fields which are not strong enough for the onset of intense cavitation are now used with success in electrode processes.

However, Yeager [15] notes that low-intensity ultrasonic waves have little effect on electrodeposition, and its effect in this case is similar to that of mixing. The presence of even insignificant cavitation in the electrolyte is effective in decreasing the limits imposed on diffusion next to the electrode. This must be taken into account when utilizing the depolarizing action of ultrasonic vibrations.

The cavitation effect is a complex combination of phenomena characterized by periodic jump-like changes in the sound intensity and temperature, resulting from the existence of two types of gas bubble vibrations in the liquid. Many investigators have neglected this phenomenon altogether and do not describe the characteristics of the field used in their experiments.

The intensity of ultrasonic vibrations cannot be measured by cavitation phenomena, since the cavitation threshold is determined not only by the intensity of the field but also by its frequency, the temperature and type of electrolyte, etc.

A. M. Ginburg [16] recommends determining the cavitation threshold by the change in the waveform of ultrasonic vibrations measured with a piezoelectric probe. Under precavitation conditions these vibrations are sinusoidal. In the presence of cavitation this shape is deformed and an edge component appears in the spectrum.

Periodic variations of the potential of the polarized electrode in a sound field were first noted by Nikitin [23] and were called sono-electrochemical phenomena. The investigation of this effect in [24] showed that, depending on the experimental conditions, the occurrence of sono-electrochemical phenomena may be related to the influence of sound waves with changes in the electric double layer, the true polarization current density, and the electrical conductivity of the solution in the electrode layer. The depolarizing action of ultrasonic vibrations during the process of electrolytic liberation of hydrogen was studied in [52]. According to [52], the depolarizing action of cavitation is due not only to energetic mixing and desorption but also to partial dehydration of ions under the effect of shock waves resulting from the closing of cavitation voids. Trofimov [32, 33] also obtained data on the effect of ultrasound on the concentration, chemical, and ohmic polarization. He concluded that, depending on the geometric parameters of the cathode, the effect of ultrasound on electrolysis may render the distribution of the metal on the electrode either more or less uniform (the degree of the effect of ultrasound on the uniformity of the coating depends on the nature of the polarization). In the case where the electrodeposition process is limited to concentration polarization, the use of ultrasound makes it possible to obtain more uniform coatings. In the case of chemical polarization its action is less effective.

Electrocrystallization and anodic dissolution of metals in an ultrasonic field are investigated in the following chapters.

ELECTRODEPOSITION OF METALS IN AN ULTRASONIC FIELD

In spite of a large number of investigations concerning the mechanism of the effect of an ultrasonic field on the electrodeposition of metals, there is still no general theory of this process. Most investigators assume that the main effect of ultrasound is similar to mechanical mixing. At the same time, other investigators have ascertained that ultrasonic vibrations induce a uniform composition of the electrolyte which cannot be attained by any other method used in electrodeposition of metals. This uniformity of the composition is particularly clear in the case of electrodeposition of metals on machine parts with complex shapes. The use of ultrasonic vibrations results in uniform coatings in cases where mechanical mixing does not provide a uniform coating.

An analysis of the experimental results shows that ultrasonic vibration changes the concentration polarization, the pH of the electrolyte, the role of hydrogen liberation in electrode processes, and also affects the structure of the electrolytic deposit. Thus, ultrasonic vibration affects all stages of the electrode processes during electrocrystallization of metals. Let us examine the known data concerning this process.

Effect of Ultrasonic Field on the Electrode Potentials
and Concentrational Polarization

Ultrasonic vibration induces intense mixing and therefore affects the value of the electrode potential during electrodeposition of metals.

The experiments in [12] showed that the ultrasonic field does not affect the equilibrium potential of the electrode. The overvoltage of the electrode resulting from polarization is much lower in the ultrasonic field. It was shown that the decrease of overvoltage under the effect of ultrasonic field is always greater than that resulting from ordinary mixing. Apparently, ultrasonic vibration not only renders the composition of the electrolyte at the electrodes more uniform, but has a definite effect on the removal of the gaseous phase from the electrolyte and from the surface of the electrodes. It was thought in [12] that cavitation might also induce a small decrease in the overvoltage.

It was noted in [25] that ultrasonic and sonic vibrations affect the equilibrium potential of the electrodes with the current off.

The effect of vibration on polarization and the study of the changes resulting from it were studied during deposition of copper from sulfates and chlorides and the deposition of zinc from sulfate, chloride, and ammonium electrolytes containing different amounts of zinc and other components (boric acid, sulfuric acid, sodium sulfate, glue).

The results of these experiments showed that, contrary to established opinion, the metal separation overvoltage in ultrasonic and sonic traveling-wave fields can be either lower or higher than without a field. The direction and magnitude of the change in polarization depend on the composition of the electrolyte, the nature and the concentration of metal ions, the current density, and the nature of surface-active additive elements and neutral ions.

At high current densities, at which the concentration limitations are considerable, the vibration induces depolarization during the deposition of copper and zinc from all the electrolytes investigated. The higher the concentration limitations, the higher the depolarization.

At low current densities, where the concentration limitations are low, vibration does not change the potentials at which copper and zinc are separated from thoroughly purified electrolytes, but the vibration increases the overvoltage to 15-20 mV in the case of separation of zinc from electrolytes which are not additionally purified and also from electrolytes containing surface-active substances.

TABLE 3

Deposition time after changing experimental conditions, min	Zinc separation overvoltage, mV	
	On a stationary cathode	On a vibrating cathode
0	61	45
0.2	54	58
1	51	60
2	49	61
5	46	61
10	45	61

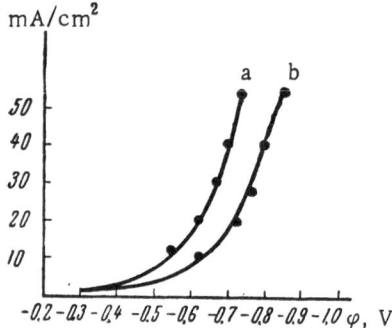

Fig. 1. Polarization curves obtained during electrodeposition of nickel. a) In an ultrasonic field; b) under ordinary conditions.

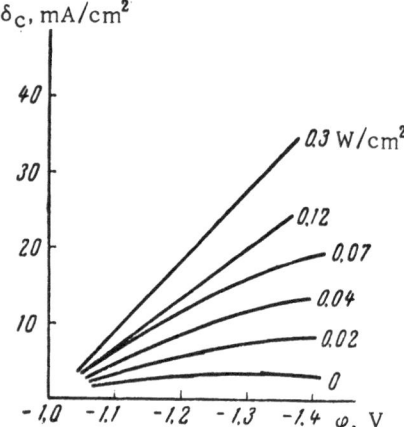

Fig. 2. Dependence of the cathodic polarization during electrodeposition of nickel on the ultrasonic field intensity.

Comparison of the effects of vibrations with different parameters showed that sonic vibrations ($f = 100$ Hz) transmitted through the electrolyte change the deposition overvoltage very little. Sonic vibrations applied directly to the cathode have a much greater effect — almost equal to that of low-frequency ultrasonic vibrations. This result refers to the decrease in overvoltage at high current densities and also to the increase in polarization at low current densities.

To clarify the nature of the change of polarization at low current densities, we made two series of experiments in which we measured the change of potential with time: a) after the polarizing current was interrupted (with a vibrating and stationary cathode); and b) after turning on and turning off the vibrations at constant current density.

A comparison of the curves representing the variation of potential with time shows that vibration increases the passivation rate of the surface.

Passivation is intensified as the interrupt time of the polarizing current is increased and reaches saturation after a 15 min interruption.

Under the effect of vibration, passivation increases particularly when slow-acting passivators are used.

The shape of the curve representing the variation of the potential with time (after the jump) is different in the presence and the absence of vibration. The difference is due not only to different values of the concentration limitations but also to the increase of the passivation of the surface of the cathode under the effect of applied vibration.

The potential either increases or decreases by jumps at the moment the vibrations are applied or interrupted. After this, the overvoltage increases very slowly when the vibrations are applied and decreases very slowly when the vibrations are interrupted.

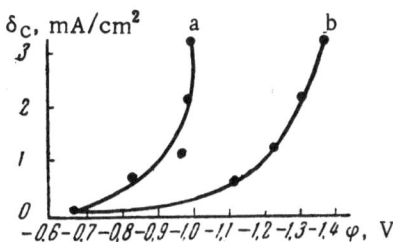

Fig. 3. Dependence of cathodic polarization on the current density during electrodeposition of tin. a) In an ultrasonic field; b) under ordinary conditions.

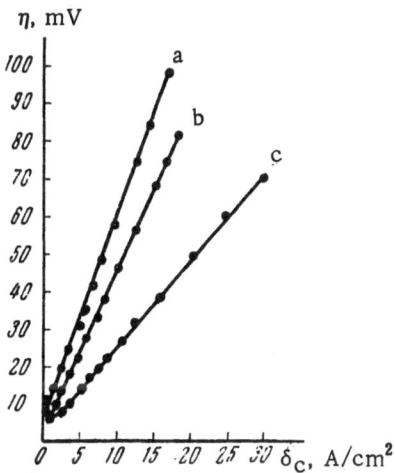

Fig. 4. Polarization curves obtained during electrodeposition of copper. a) Under ordinary conditions; b) in a continuous-flow electrolyte; c) in an ultrasonic field.

The rapid jump of the potential at high current densities and low metal ion concentrations corresponds either to the elimination or creation of concentration polarization. The slower changes of the potential are explained by the change in the real surface on which deposition occurs, i.e., "accommodation" to the deposition conditions characterized by new conditions of the diffusion of metal ions and of the passivation of the cathode surface. At low current densities and high zinc ion concentrations in solutions not subjected to additional purification the overvoltage increases by a jump when vibration is applied and decreases by a jump when the vibration is stopped. The typical variation of the potential with time under these conditions in the case of deposition of zinc from an electrolyte not additionally purified and containing 200 kg/m^3 ZnSO$_4 \cdot$7H$_2$O at a current density $\delta_c = 100$ A/m^2 is shown in Table 3.

In electrolytes additionally purified, there are no potential jumps characteristic of low current densities and high concentrations of zinc ions. The jumps occur, however, when $2 \cdot 10^{-6}$ kg/m^3 of joiner's glue is added to the solution. All these results allow us to explain the increase of the zinc deposition overvoltage under the influence of vibration by an increase of the passivation rate of the cathode surface.

At low current densities, when the concentration limitations are small, vibration increases the overvoltage due to the increased passivation rate of the cathode surface.

At high current densities, when the concentration polarization decreases much more due to vibration than the overvoltage increases due to passivation, the overvoltage decreases (depolarization). Many investigators have shown that the polarization of nickel, copper, tin, silver, chromium, iron, and other metals decreases under the influence of the ultrasonic field [10-12, 16, 25-29].

The effect of the ultrasonic field is particularly great in the case of electrocrystallization of nickel. According to the results in [47], ultrasonic vibration decreases the polarization by more than 100 mV (Fig. 1). In [21] it was also found that the potential for the deposition of nickel decreases considerably in an acoustic field created by vibration with a frequency of 34 kc and an intensity of $3 \cdot 10^3$ W/m^2. The author showed that the effect of the ultrasonic field decreases with increasing intensity and frequency of the vibration (Fig. 2). In [47] the authors measured the polarization curves during electrodeposition of tin (Fig. 3) The results showed that tin precipitates from stannates at considerably lower potentials in an ultrasonic field.

The effect of ultrasonic vibration on the separation potential of copper has been investigated by many authors [11, 30-34]. Figure 4 shows the polarization curves obtained during electrodeposition of copper from a solution containing 0.17 kmole/m^3 of CuSO$_4$ and 0.02 kmole/m^3 of H$_2$SO$_4$. It was found that ultrasonic vibration applied to the electrolyte has an immediate effect. The potential drops so rapidly that it is impossible to obtain experimental points at the beginning of the curve. The drop of the potential is so rapid because of the elimination of the concentration gradient by ultrasonic vibration, and its subsequent decrease is due to the change in the effective surface area of the cathode. The increase in polarization is sharper at low concentrations. It is interesting to note that the effect of ultrasonic vibration is much greater in solutions saturated with gas.

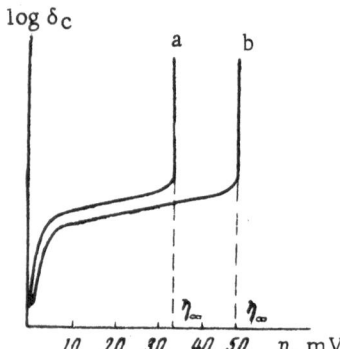

Fig. 5. Effect of ultrasonic vibration on the overvoltage during deposition of copper. η) Overvoltage during deposition of copper; η_∞) limiting overvoltage; a) in ultrasonic field; b) under ordinary conditions.

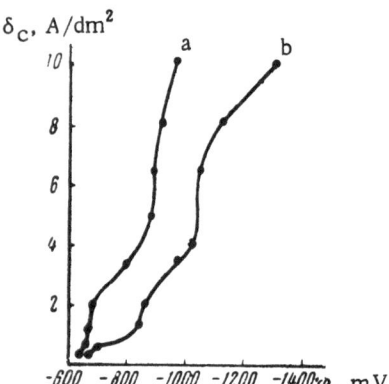

Fig. 6. Polarization curves obtained during electrodeposition of silver. a) Under ordinary conditions; b) in an ultrasonic field.

Probably the cavitation effect plays a role in this case. It was found in [31] that if acoustic vibrations are applied to the electrolyte the overvoltage of the copper cathode decreases considerably. Figure 5 shows the variation of the separation potential of copper with the logarithm of the current density. The effect of ultrasonic vibration on cathodic polarization during the deposition of silver was studied in [21]. Figure 6 shows that the separation potential of silver from cyanide decreases in an ultrasonic field. It was shown in [27] that ultrasonic vibration decreases cathodic polarization during electrodeposition of chromium. Figure 7 shows that ultrasonic vibration has an effect only before the precipitation of chromium; after the deposition of chromium begins, the ultrasonic field has no significant effect on the value of the cathode potential.

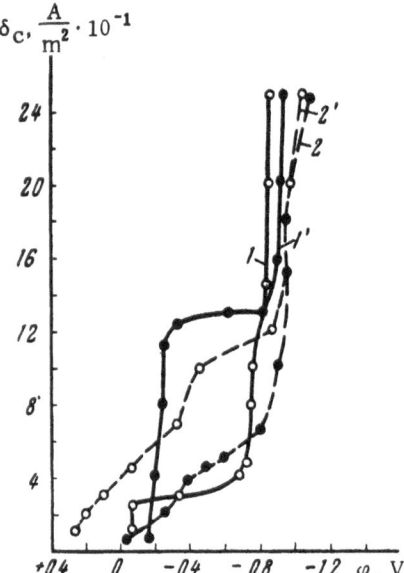

Fig. 7. Polarization curves obtained during electrodeposition of chromium. 1, 2) Without ultrasonic field; 1', 2') in ultrasonic field. Cathodes 1 and 1' are made of iron; cathodes 2 and 2' are made of chromium.

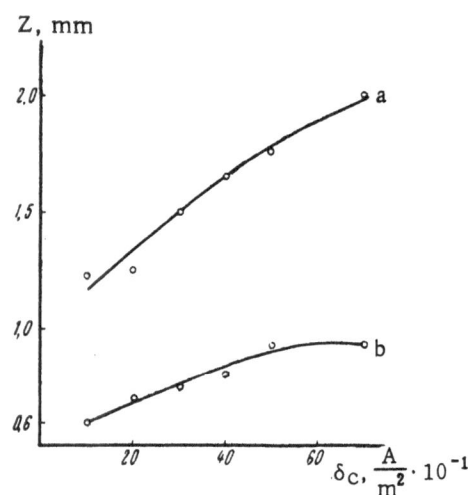

Fig. 8. Variation of internal stresses in nickel deposits under the influence of an ultrasonic field. a) Under ordinary conditions; b) in an ultrasonic field.

TABLE 4. Effect of an Ultrasonic Field on the Hydrogen Liberation Potential (According to Data in [53])

Cathode material	Cathode potential in weak ultrasonic field, mV	Cathode potential in strong ultrasonic field, mV	Decrease of cathode potential in ultrasonic field, mV
Platinum	900	400	500
Copper	1250	320	930
Brass	1100	320	780
Iron	1140	460	680
Nickel	1140	360	780
Tin	1320	500	820
Lead	1320	360	960
Zinc	1640	780	860
Aluminum	1320	600	720
Magnesium	1540	1640	−100

Effect of an Ultrasonic Field on the Evolution of Hydrogen

It is well known that the electrodeposition of some metals is accompanied by the simultaneous discharge of hydrogen, which induces considerable changes in the electrocrystallization process.

Some authors have shown that the hydrogen liberation potential decreases considerably under the effect of an ultrasonic field during electrolysis [13, 15, 18]. The characteristic effect of an ultrasonic field on the hydrogen liberation potential from a 0.4 N Na_2SO_4 solution at a current density of 25 A/m^2 is shown in Table 4.

In [13] the authors investigated the depolarizing effect of ultrasonic vibrations on the evolution of hydrogen during electrolysis. They found that ultrasonic vibrations of 5000 W/m^2 at a frequency of 21.3 kc decrease the electrode potential by 450 mV in the case of hydrogen liberation from a 0.5 N KOH solution, the current density varying from 314 to 945 A/m^2.

An increase in the temperature does not change the depolarization effect, which remains the same at 20-60°C within the range of current densities indicated.

The authors in [14] also investigated the variation of the oxygen polarization of a smooth platinum electrode, an electrode coated with lead dioxide, and an electrode coated with oxidized lead under the influence of ultrasonic vibration. They showed that ultrasonic vibration decreases polarization during the evolution of oxygen on smooth platinum. The oxygen overvoltage decreases also on a platinum electrode coated with lead dioxide. The change in polarization due to the evolution of hydrogen during electrolytic deposition of nickel showed that the hydrogen liberation potential and metal separation potential decrease and that the degree to which the potentials decrease is smaller as the frequency is increased.

Experimental results obtained at the Kazan' Chemicotechnological Institute showed that ultrasonic vibrations greatly decrease the internal stresses in the deposit resulting from hydrogen absorption.* At the same time, the stresses become more uniform over the whole surface of the electrode. For example, in the case of electrolytic deposition of nickel coatings 20-30 microns thick the bending distance (distance from the vertical) of the cathode plate on which nickel was deposited in an ultrasonic field is 0.90 mm, as compared to 2.00 mm for plates coated under ordinary conditions. These results are shown in Fig. 8.

*For more details see p. 25.

TABLE 5. Dependence of the Depolarization Effect on the Current Density and Temperature (Fe Electrode in 0.5 N KOH Solution)

Current density, A/m²·10⁻¹	20°C			30°C			40°C			50°C			60°C		
	φ_{ord}	φ_u	$\Delta\varphi$	φ_{ord}	φ_u	$\Delta\varphi$	φ_{ord}	φ_u	$\Delta\varphi$	φ_{ord}	φ_u	$\Delta\varphi$	φ_{ord}	φ_u	$\Delta\varphi$
3.14	1360	900	460	1320	870	450	1290	830	460	1260	810	450	1240	800	440
6.3	1440	990	450	1380	920	460	1360	910	450	1320	870	450	1300	850	450
9.45	1470	1340	130	1410	1260	150	1390	940	450	1370	920	450	1350	900	450
12.7	1510	1410	100	1430	1300	130	1410	1250	160	1390	1220	170	1380	1210	170
15.7	1520	1470	50	1440	1340	100	1440	1320	120	1420	1300	120	1400	1270	130

Note: φ_{ord}) Potential of polarized electrode, mV; φ_u) potential in ultrasonic field, mV; $\Delta\varphi$) depolarization effect, mV.

The magnitude of the internal stresses is proportional to the bending distance of the cathode Z:

$$\tau = \frac{1}{3}\frac{Ea^2}{l^2\sigma}Z,$$

where τ is the internal stress; E is the Young's modulus; d is the thickness of the base; l is the length of the cathode; and σ is the thickness of the substrate.

Increase of the Deposition Rate of Metals in an Ultrasonic Field

In a number of cases electrodeposition can be carried out at high current density because of the decrease in polarization. This is due to the fact that the composition of the layer of electrolyte adjacent to the electrode is rendered more uniform. For example, good quality nickel deposits can be obtained at current densities 3-4 times higher in an ultrasonic field than under ordinary conditions. In [35-37] these results were checked industrially and showed that the use of ultrasonic vibrations with a frequency of 18-20 kc greatly accelerates the deposition of copper, zinc, cadmium, and tin. Table 6 shows the possible increase in current density with ultrasonic vibration. Investigations in [30] of the electrodeposition of copper, brass, and silver from cyanide electrolytes in an ultrasonic field showed that the use of ultrasonic vibration allows one to use a current density 5-6 times higher than those used in ordinary electrodeposition in the case of copper plating (Fig. 9).

The highest acceptable current density is the current density at which the deposits are of good quality. An increase of the current density beyond this value leads to the formation of dendrites. Figure 10 shows that the rate of copper plating from cyanide solutions can be increased by a factor of 15-20 in an ultrasonic field. The study of the dependence of the yield on the current density at different temperatures showed that the yield is approximately 94-98% at temperatures between 20 and 50°C. It was found that acoustic vibration is most effective when the concentration of copper cyanide is about 80 kg/m³. The range of current densities used in brass plating is broadened considerably in an ultrasonic field. The increase in the acceptable current density and the increase in the yield make it possible to increase the rate of deposition of brass to 120-150 μ/h, which is 20-30 times higher than the ordinary rate. Acoustic vibration changes the composition of the brass in the coating. For example, when the current density is 20 A/m² the brass deposited without

TABLE 6. Effect of an Ultrasonic Field on the Limiting Permissible Current Density

Metal	Composition of the electrolyte	Amount of substance per m^3, kg/m^3	Current density, A/m$^2 \cdot 10^{-2}$		Source of data
			Under ordinary conditions	In an ultrasonic field	
Zinc	$ZnSO_4$ $Al_2(SO_4)_3$ Na_2SO_4	215 204 100	Limit	20	[29]
	ZnO $NaOH$ $NaCN$	50 120 120	Limit	30	[37]
Chromium	Acid electrolyte	—	Limit	3-5 times higher than under ordinary conditions	[59]
Cadmium	CdO $NaCN$ Na_2SO_4 $NiSO_4$	30-50 130 30-60 0.8	0.5-0.6	1.8	[37]
	Cyanide electrolyte	—	27	81	[59]
Nickel	$NiSO_4$ $C_3H_4(OH) \cdot (COONa)_3$	40 35	0.3	4	[21]
	Cyanide electrolyte	—	Limit	15 times higher	[59]
	$NiSO_4$ H_3BO_3 pH	170 26 5.38	Limit	7.5 times higher	[29]
	Acid electrolyte	—	Limit	8	[37]
	$NiSO_4$ H_3BO_3 $NaCl$ NaF Na_2SO_4	250 25 15 10 230	1.6	40	[36]
	Ni H_3BO_3 Cl^- $C_{10}H_6(HSO_4)_2$ NaF pH	44.8 33.8 18.4 0.8 4.8 5.8	1.6	40	[36]
	Acid electrolyte	—	Limit	14	[38]

TABLE 6 (Continued)

Metal	Composition of the electrolyte	Amount of substance per m³ kg/m³	Current density, A/m² · 10⁻²		Source of data
			Under ordinary conditions	In an ultrasonic field	
Copper	$CuSO_4$ H_2SO_4 C_2H_5OH	200 50 10	Ordinary	15	[57]
	Cyanide electrolyte	—	Ordinary	20-30 times higher	[59]
	$Cu(CN)_2$ NaCN $(NH_4)_3PO_4$	60 20 5	1	8 times higher	[37]
	$CuSO_4$ H_2SO_4	200 60	5	16	[37]
	$CuSO_4$ H_2SO_4	250 75	15	125	[35]
	$CuSO_4$ $Na_4P_2O_7$ Na_2HPO_4	70 200 95	Ordinary	8	[36]
Silver	$K(Ag(CN)_2)$ KCN K_2CO_3	40.2 38.3 89.3	3	10-20	[30]

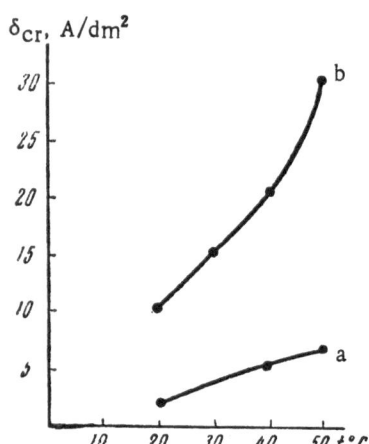

Fig. 9. Dependence of the critical current density on the temperature during electrodeposition from a solution containing 120 kg/m³ $Cu(CN)_2$. a) Under ordinary conditions; b) in ultrasonic field.

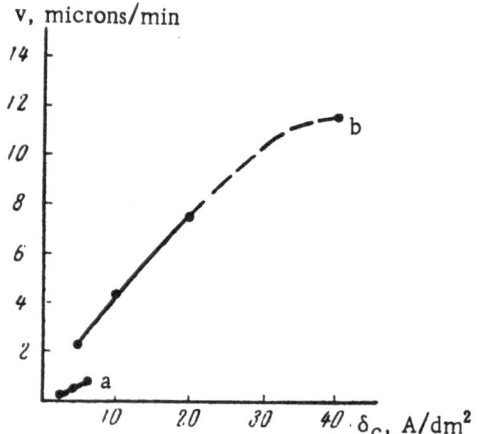

Fig. 10. Dependence of the deposition rate of copper on the current density. a) Without ultrasonic field; b) with ultrasonic field.

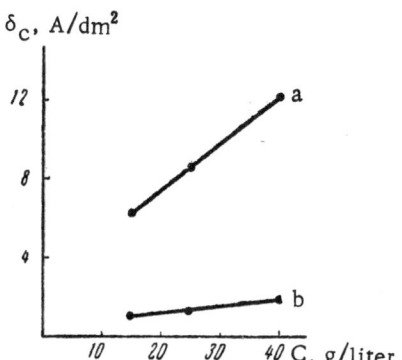

Fig. 11. Dependence of the critical current density on the concentration of silver in the electrolyte at 40°C. a) In ultrasonic field; b) under ordinary conditions.

ultrasonic vibration contains 81% copper, while it contains 68% copper when ultrasonic vibration is used. This is apprently due to the fact that ultrasonic vibration affects the polarization of copper and zinc differently. The working current densities used in silver plating can be increased to 120 A/m² when ultrasonic vibration is used, and the rate of deposition of silver is 6-7 μ/min. Figure 11 shows the dependence of the critical current density on the concentration of silver in the electrolyte. It was found that the yield of the metal does not change, remaining at 85-96%. Ultrasonic vibration has little effect on the mixing of the electrolytes used for copper plating, brass plating, and silver plating, and the mixing decreases with increasing current density. The dependence of the acceptable current density on the intensity of ultrasonic vibration at different temperatures was shown in [29]. The results of this investigation are shown in Fig. 12. The critical current density is represented along the y axis and the intensity of ultrasonic vibration along the x axis. The same authors investigated the dependence of the amount of metal precipitated per unit time per unit surface P on the intensity of the field I (Fig. 13).

Table 7 shows the degree to which the electrodeposition of metals is accelerated by the application of an ultrasonic field. The authors in [35] succeeded in increasing the current density by almost a factor of eight in the case of electrolytic plating of wire with copper. Their results are given in Table 8.

The high rate of mixing induced by ultrasonic vibration decreases the concentration polarization and makes it possible to obtain electrolytic deposits from electrolytes with much lower concentrations of salts than those usually used in electrolytic plating. Thus, electrolytic deposits of copper were obtained in [35] with current densities of 600-1000 A/m² from very dilute copper sulfate solutions. Similar results were obtained recently in [34]. A. I. Sobolev noted that good quality nickel coatings can be obtained with low concentrations of nickel salts in the electrolyte [34, 36, 38]. This author also showed that with ultrasonic fields one can use electrolytes containing considerable amounts of slime. The same results were obtained in the laboratory of the Kazan Chemicotechnical Institute.

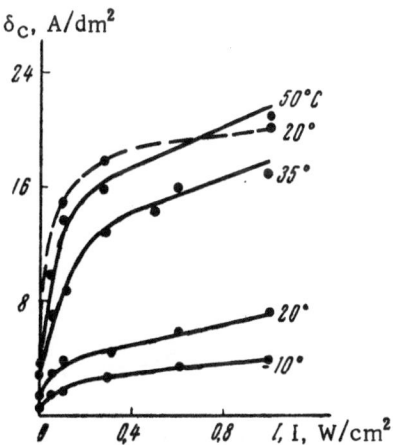

Fig. 12. Dependence of the limiting current density on the ultrasonic field intensity in the case of electrodeposition of nickel (solid line) and zinc (dashed line).

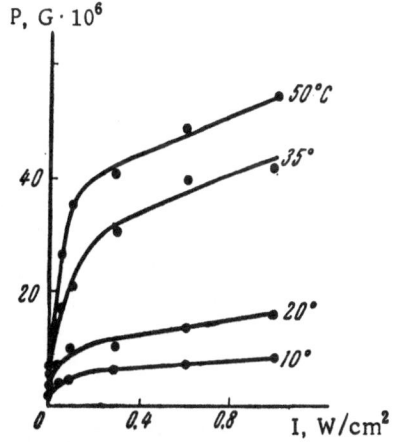

Fig. 13. Dependence of the amount of nickel deposited on the ultrasonic field intensity at different temperatures.

TABLE 7. Effect of the Ultrasonic Field Intensity and Temperature
of the Electrolyte on the Rate of Deposition of Copper

Ultrasonic field intensity, $W/m^2 \cdot 10^{-4}$	Process accelerated by a factor of:			
	10°C	20°C	35°C	50°C
0	1	1	1	1
0.05	2.8	1.85	2.61	3.72
0.1	3.5	2.64	3.06	4.93
0.3	4.6	2.7	4.7	5.7
0.6	5.28	3.5	5.9	6.6
1.0	5.92	4.12	6.28	7.36

TABLE 8. Conditions of Electrolytic Copper Plating of Wire

Conditions of electrolyte	Deposition conditions		Thickness of deposit, μ
	Current density, A/m^2	Deposition time of copper, min	
In a stationary electrolyte	1500	30	130
In an electrolyte vibrating with a frequency of 100 Hz	7500	5	170
Mechanically mixed electrolyte	5000	10	200-250
Electrolyte in ultrasonic field	12500	5	200

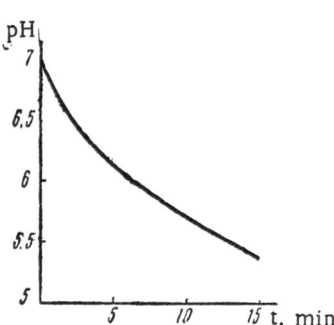

Fig. 14. Effect of ultrasonic
vibration on the acidity of the
water. The ultrasonic intensity
is $I = 3 \cdot 10^4$ W/m², the frequency is $f = 800$ kc.

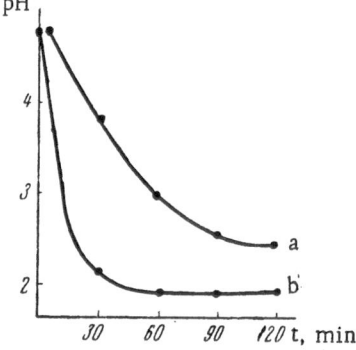

Fig. 15. Variation of the acidity
of the electrolyte containing 280
kg/m³ $NiSO_4 \cdot 7H_2O$ under the effect
of ultrasonic vibration. a) Under ordinary conditions; b) in ultrasonic field.

TABLE 9. Amounts of Hydrogen and Oxygen Evolved During Electrolysis

Current density, mA/cm²	Amounts of electrolysis products under ordinary conditions, cc		Amounts of electrolysis products in ultrasonic field, cc		Change in the amounts of electrolysis products, %	
	H_2	O_2	H_2	O_2	H_2	O_2
20	3.7	1.5	4.2	1.8	13.0	20
40	7.5	3.8	8.6	4.2	14.6	10
60	11.5	5.4	12.5	6.0	11.3	11
80	15.2	7.2	16.8	8.4	10.5	14

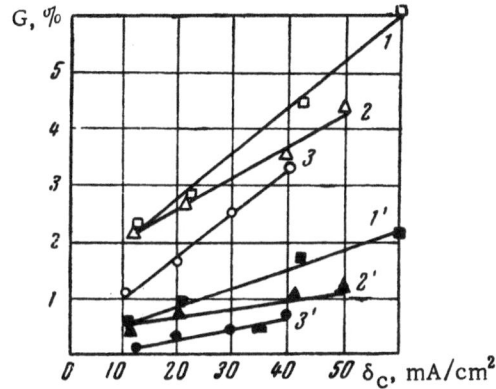

Fig. 16. Effect of an ultrasonic field on the gas impregnation of the metal (G, %) during electrolysis. 1) Under ordinary conditions (10% H_2SO_4, H_2); 1') same, in ultrasonic field; 2) under ordinary conditions (17% NaOH, H_2); 2') same, in ultrasonic field; 3) under ordinary conditions (26% NaCl, Cl_2); 3') same, in ultrasonic field.

It is important to note that with an ultrasonic field one can use solutions of simple salts without adding buffers or compounds to increase their electrical conductivity. Under the influence of an acoustic field one can obtain electrolytic deposits of nickel from $NiSO_4 \cdot 7H_2O$ alone, and no significant change in the acidity of the solution occurs during electrodeposition.

Thus, all these results show that the technique of electrolytic deposition of metals can be simplified and control of the composition of the electrolyte facilitated by the application of ultrasonic fields.

Changes Occurring in the Electrolyte As the Result of Ultrasonic Treatment

Change of the pH and the Electrical Conductivity of the Electrolyte

The pH of the water decreases from 7 to 5.6 after 15 min of vibration with an ultrasonic field with an intensity of $3 \cdot 10^4$ W/m² and a frequency of 800 kc. The reason for this decrease of the pH is the conversion of nitrogen into nitric acid as the result of ultrasonic vibration. The presence of nitric acid was demonstrated by chemical analysis. Applications of the ultrasonic field longer than 15 min do not change the pH of the water any further (Fig. 14) [39]. A study of the pH variation during electrodeposition of nickel by G. Ya. Vyaseleva showed that ultrasonic vibration maintains the pH at a given level (Fig. 15). In our experiments on electrodeposition of nickel the alkalinity of the near-cathode layer was much lower.

Thus, by using ultrasonic vibration one can deposit metallic nickel from solutions without buffers and under conditions in which the whole cathode becomes covered with hydroxides in the absence of ultrasonic vibration. It should be emphasized that the electrical conductivity of water and of electrolytes changes under the influence of ultrasonic vibration [45-50]. For example, the specific electrical conductivity of water increases from $2.0 \cdot 10^{-4}$ $\Omega^{-1} \cdot$ m^{-1} to $3.5 \cdot 10^{-4}$ $\Omega^{-1} \cdot$ m^{-1} and the electrical conductivity of potassium chloride solutions increases with decreasing concentrations.

It was found in [40] that the electrical conductivity of electrolytes subjected to ultrasonic irradiation increases to a certain maximum and then remains constant. The electrical conductivity also varies as the result of elimination of the gaseous phase [41].

In this respect the experiments on degassing of electrolytes during the preparation of hydrogen, oxygen, and chlorine are particularly characteristic.

TABLE 10. Effect of Ultrasonic Field on the Evolution of Chlorine and Hydrogen

Temp., °C	Amounts of electrolysis products under ordinary conditions, cc		Amounts of electrolysis products after ultrasonic treatment of electrolyte, cc		Amounts of electrolysis products in ultrasonic field, cc		Changes in the amounts of electrolysis products, %	
	H_2	Cl_2	H_2	Cl_2	H_2	Cl_2	H_2	Cl_2
20	8	5	0.1	5.0	9	11	12	120
40	8	6	0.1	4.0	9	11	12	83
60	8	7	0.1	2.0	9	11	12	57

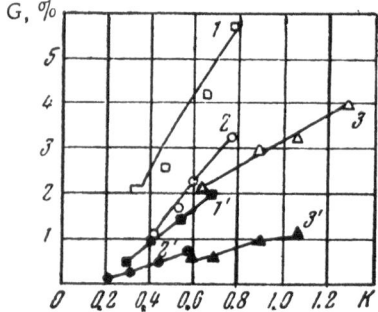

Fig. 17. Variation of the resistance factor K of the electrolyte as a function of the amount of gas in the metal. 1) Under ordinary conditions (10% H_2SO_4, H_2); 1') same, in ultrasonic field; 2) under ordinary conditions (17% NaOH, N_2); 2') same, in ultrasonic field; 3) under ordinary conditions (26% NaCl, Cl_2); 3') same, in ultrasonic field.

Hydrogen and oxygen were obtained by electrolysis of 10% H_2SO_4 and 17% NaOH solutions. The electrolysis was carried out in a specially designed vessel. The amount of the products of electrolysis was determined by the difference in levels before and after electrolysis.

The gases evolved during electrolysis saturate the electrolyte and increase its electrical resistance. As the result, the electrolyte is filled with gas in amounts which can be calculated by the following formula:

$$G = \frac{\Delta h}{h} \cdot 100\%,$$

where h is the height of the column of liquid after it is degassed; Δh is the decrease in the liquid level resulting from degassing.

The electrical resistance of the electrolyte, which depends on the amount of gas in the electrolyte, was measured by the ordinary compensation method with platinum electrodes soldered to the walls of the vessel.

Table 9 shows the data on the amounts of decomposition products from a 10% H_2SO_4 solution under ordinary conditions and in the presence of an ultrasonic field. The investigation was made at 20°C; the time of electrolysis was 10 min.

The data show that in all cases ultrasonic vibration affects the yield of the products of electrolysis. The amount of hydrogen evolved during decomposition of a 10% H_2SO_4 solution and a 26% NaOH solution increases by an average of 11-12%. Ultrasonic vibration has a particularly appreciable effect on the evolution of chlorine — the yield increases by almost 40%. Such a change in the yield of the products of electrolysis in an ultrasonic field is apparently due to the decrease in liberation potential of the gas. The depolarization jump is particularly high for hydrogen. It increases with increasing current density, reaching almost 300 mV.

Degassing of the electrolyte plays a very important role in the increase of the yield of gaseous products of electrolysis. In an ultrasonic field the gas bubbles have no time to dissolve in the electrolyte; they float rapidly to the surface, thus increasing the yield of the products of electrolysis. The effect of degassing is particularly important during the first 10-15 min of electrolysis. After this period the difference in the amount of chlorine or hydrogen obtained in an ultrasonic field and under ordinary conditions decreases. The effect on gas evolution also decreases as the temperature of the electrolyte is increased. Data on the dependence of the amounts of chlorine and hydrogen produced during electrolysis on the temperature of the electrolyte are given in Table 10.

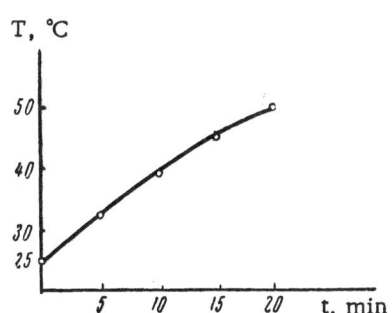

Fig. 18. Effect of ultrasonic vibration on the temperature of an electrolyte containing 280 kg/m^3 NiSO$_4 \cdot$ 7H$_2$O.

The solubility of hydrogen varies with the temperature of the solution much less than the solubility of chlorine. With increasing temperatures between 20 and 60°C the solubility of chlorine decreases and the increase in the yield due to vibration decreases from 12 to 5.7%. It is interesting to note that if one subjects the electrolyte to ultrasonic vibration immediately after ordinary electrolysis, the gas dissolved in the electrolyte begins to evolve. If one subjects the electrolyte to ultrasonic vibration after electrolysis with an ultrasonic field, no significant amount of gas is evolved, regardless of the temperature.

Thus, the degree of degassing and the yield of the products of electrolysis in an ultrasonic field depend greatly on the time of electrolysis and the temperature of the electrolyte.

The electrical resistance of the electrolyte changes with the amount of gas contained in it. Figures 16 and 17 show data on the amount of gases and the electrical resistance of H$_2$SO$_4$, NaOH, and NaCl electrolytes under ordinary conditions and with an ultrasonic field. The amount of gas in the electrolyte was determined as the sum of the amounts of hydrogen and oxygen in the H$_2$SO$_4$ and NaOH electrolytes and the sum of the amounts of hydrogen and chlorine in the NaCl electrolyte. The curves show that the amount of gas in the electrolyte subjected to ultrasonic vibration decreases sharply — by almost three fold. The ohmic resistance of a degassed 10% H$_2$SO$_4$ solution decreases by a factor of 1/1.5, that of the 26% NaCl solution decreases by a factor of 1/1.2, and that of a 17% NaOH solution decreases by a factor of 1/1.3.

The degassing of liquids in an ultrasonic field with an intensity of $2 \cdot 10^4$ W/m^2 was investigated in [42]. The author came to the conclusion that gases are eliminated in two stages: a) the gas bubbles unite into complexes; b) the complexes are occluded to the surface of the liquid.

The time and the degree of degassing depend greatly on the ultrasonic intensity.

Cavitation, which produces a "sucking" action, plays an important part in this process. In the absence of cavitation small bubbles are gathered at the crest of the wave by the sound pressure and this facilitates the occlusion of the bubbles to the surface, resulting in degassing of the liquid.

It was found that ultrasonic vibration decreases the amount of gas in the electrolyte and, as a result, the ohmic resistance of the electrolyte decreases during electrolysis. This decrease in the electrical resistance lowers the expenditure of electrical energy per unit product of electrolysis. The study of the effect of ultrasonic vibration on electrolytic separation of deuterium showed that the gas is rapidly evolved from a polished platinum electrode. It was assumed that this effect prevents a catalytic type of reaction between the discharging hydrogen and the electrolyte. As a result, the separation coefficient of deuterium increases.

This occurs during electrocrystallization of nickel, iron, cobalt, and other metals whose precipitation is accompanied by the evolution of large amounts of hydrogen. The change in the electrical conductivity of the electrolyte is the result of the ultrasonic vibrations. Ultrasonic vibrations affect not only the pH value and the volume of gas evolved but also the temperature and chemical reactions which occur.

Ultrasonic vibration raises the temperature of the electrolyte quite rapidly. In the case of nickel plating with a current density of 400 A/m^2 from an electrolyte containing 280 kg/m^3 NiSO$_4 \cdot$ 7H$_2$O the temperature of the electrolyte increases rapidly, and after 10 min of electrolysis the increase is 15°C when the intensity of ultrasonic vibration is $3.5 \cdot 10^4$ W/m^2 and the frequency is 28 kc (Fig. 18). The temperature fluctuation is particularly large at the electrode-electrolyte boundary, where the implosion of cavitation voids strongly affects the temperature of the near-electrode layer. Therefore, the electrolytes should be cooled in cases where the process is more efficient at room temperature, although, because of cavitation, it is impossible to create an absolutely uniform temperature.

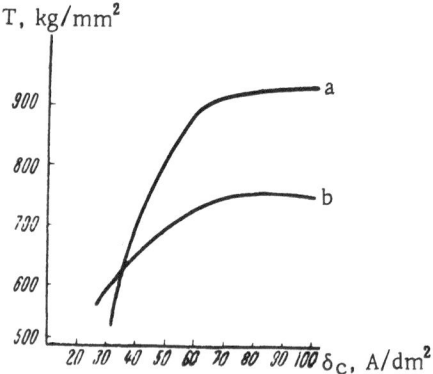

Fig. 19. Effect of ultrasonic vibration on the microhardness of chromium deposits. a) In ultrasonic field; b) under ordinary conditions.

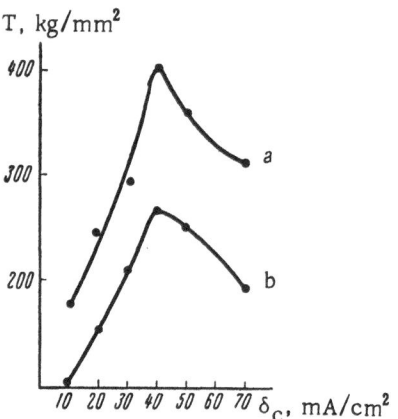

Fig. 20. Effect of ultrasonic vibration on the microhardness of nickel deposits (ultrasonic frequency 23 kc; intensity $2 \cdot 10^4$ W/m^2; t = 30 min; temperature 30°C; the electrolyte contained 280 kg/m^3 NiSO$_4 \cdot$ 7H$_2$O). a) In ultrasonic field; b) under ordinary conditions.

Variation of the Structure and Properties of Metals Deposited in An Ultrasonic Field

Ultrasonic and sonic vibrations decrease the polarization during the electrodeposition of metals. This should produce deposits with large grains and decrease the mixing of the electrolyte. However, many experimental results indicate that this is not so — the deposits consist of small crystals and often the mixing of the electrolyte is improved.

All the investigators who have studied the properties of electrolytic deposits produced in an ultrasonic field come to the same conclusion, namely, that the physicomechanical qualities of the deposits produced in an ultrasonic field are superior to those of deposits produced without ultrasonic treatment and that ultrasonic irradiation changes the structure of the deposit. Ultrasonic vibration decreases the size of the crystals, changes the shape of the crystals, increases the hardness, decreases the porosity, and improves the brightness of the deposit. A number of investigators explain these results by the acceleration and increase of cathodic passivation. It was shown in [12] that the rate of passivation of the electrodes in an ultrasonic field is twice as high as without the field because vibration accelerates oxidation processes. The effect of acoustic vibration on the passivitiy of the cathodes during electrocrystallization of zinc was studied in [25]. The authors found that the degree and rate of passivation without current in an ultrasonic field is twice that without the ultrasonic field. The cathode is also much more passivated during electrocrystallization. The change in the magnitude and rate of passivation of the cathode under the effect of acoustic vibration is due to the change in the crystal structure of the precipitates produced in an ultrasonic field.

The change in the degree of passivity of the electrode and the change in concentration polarization determine the grain size of the deposit and affect the mixing of the electrolyte. When the degree of passivation is higher than the decrease of the concentration polarization the deposits consist of small crystals; when the situation is reversed the deposits consist of large crystals.

It is indicated in [25] that this fact is responsible for the contradictory information regarding the effect of ultrasonic vibration on the structure of electrolytic deposits.

Let us describe some experimental results. In [38] the authors found that the hardness of chromium increases 40%. The variation of the microhardness of electrolytic deposits of chromium in an ultrasonic field is shown in Fig. 19.

B. Brown showed that plating obtained in ultrasonic fields is almost free from impurities, has a high breaking strength, and is very hard. However, the findings of other investigators contradict these results.

In [57] the author indicates that chromium deposits produced in an ultrasonic field are no harder than those produced without it. In [26] and [27] it was found that ultrasonic vibration has practically no effect on the hardness of electrodeposited chromium and zinc. The microhardness of chromium deposited without ultrasonic vibration was 1145-1315 kg/mm^2 and 1225-1415 kg/mm^2 with ultrasonic vibration. A. M. Ginberg and M. A. Naishuler did not find any difference in the microhardness of deposits produced under ordinary conditions and in an ultrasonic field. The hardness of nickel deposits produced in an ultrasonic field is lower. Kozan also found that ultrasonic vibration does not significantly affect the hardness of chromium or nickel deposits. A. N. Trofimov also indicated that the microhardness of nickel deposits produced in an ultrasonic field is about 100 kg/mm^2 lower than that of nickel deposits produced under ordinary conditions.

Investigations carried out in the laboratory of the Kazan Chemicotechnological Institute showed that deposits produced in an ultrasonic field are 60% harder than those produced under ordinary conditions. The results of these experiments are shown in Fig. 20. The curves show that the variation of the hardness with a change in the current density is unaffected by ultrasonic vibration. The variation of the hardness with the current density passes through a maximum when the current density is 4 A/dm^2. The change in the microhardness of the deposits with the change in current density is due to the difference in the quality of deposits obtained at low and high current densities. Apparently, ultrasonic vibration has no effect on the singularities of crystal growth at different current densities.

The microhardness of deposits produced in an ultrasonic field and under ordinary conditions usually decreases with the temperature of the electrolyte between 30 and 60°C. Within this temperature range the microhardness of deposits produced with ultrasonic vibration is always higher than that of deposits produced under ordinary conditions, although the rate of the change in microhardness with increasing temperature is different in the two cases. For example, under ordinary conditions the microhardness decreases most appreciably between 30 and 40°C. Further increase in the temperature of the electrolyte has little effect on the microhardness, the total decrease being 80 kg/mm^2. In an ultrasonic field the microhardness does not change when the temperature of the electrolyte is raised from 30 to 50°C, but with a further increase in temperature the microhardness decreases to such a degree that it becomes similar to that of deposits produced without ultrasonic vibration.

In general, the microhardness of deposits produced in an ultrasonic field decreases by 180 kg/mm^2. The difference in the variation of the microhardness of deposits produced with and without ultrasonic vibration with the temperature of the electrolyte is due to the singularities of the process of electrodeposition of metals in an ultrasonic field. Some authors believe that the deposits become less hard with increasing temperature of the electrolyte because the amount of hydrogen included in them decreases. In an ultrasonic field the degree of hydrogenation of the deposits decreases to such an extent that small decreases in the amount of hydrogen absorbed as the result of the increase in the temperature have no significant effect on the microhardness. Therefore, the microhardness of deposits produced in an ultrasonic field does not vary over a rather wide range of temperatures.

The pH of the electrolyte has a significant effect on the microhardness of the deposits. The microhardness of nickel deposited in an ultrasonic field increases with increasing pH. Under ordinary conditions, the current density being 40 mA/cm^2 and the temperature 30°C, electrolytic deposits cannot be obtained with ordinary electrolytes with a pH higher than 3. The microhardness of these deposits is low, since the cathode layer becomes strongly alkaline under these conditions. The hydroxide formed as the result "lodges" between the growing crystals and is included in the deposit. All this leads to a decrease in microhardness of the deposit, so that there is no sharp increase in the microhardness of deposits produced under ordinary conditions with changing pH of the electrolyte. In an ultrasonic field the cathode layer does not become alkaline; therefore dense deposits are produced even when the pH of the electrolyte is high. The variation of the microhardness of deposits with the acidity of the electrolyte is in agreement with the published data.

It is well known that the increase in microhardness is the result of many factors: It may be due to the inclusion of foreign particles in the deposit, to a decrease in the size of the crystals, to an increase in the packing density of atoms in the crystal lattice, etc. Ultrasonic vibration affects these factors differently, depending on its intensity. Under certain conditions deposits consisting of large crystals are formed in the ultrasonic field, and then the microhardness is low. The same ultrasonic vibration which promotes the formation of deposits consisting of small crystals increases the microhardness of the deposit.

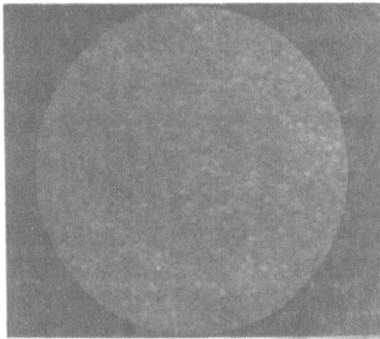

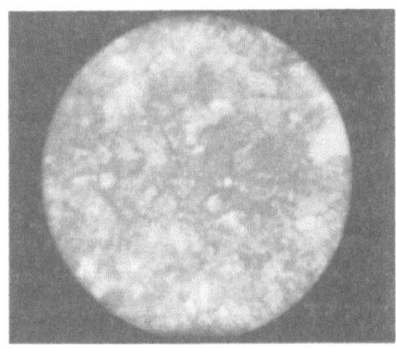

Fig. 21. Photomicrograph of the surface of a cathodic deposit obtained in an ultrasonic field.

Fig. 22. Photomicrograph of the surface of a cathodic deposit obtained under ordinary conditions.

An analysis of the results obtained in the laboratory of the Kazan Chemicotechnological Institute with respect to the effect of ultrasonic vibration on microhardness leads us to conclude that the structure of the deposit affects its mechanical properties. Thus, the most important causes of the increase in microhardness are the small crystal structure and the packing density of the crystals. The inclusions of hydrogen and other substances and the lattice deformations are not the determining factors with respect to the change in microhardness. In fact, investigations of hydrogenation and x-ray analysis have shown that the amount of hydrogen in deposits produced with ultrasonic vibration is half that in deposits produced without ultrasonic vibration. In deposits produced with ultrasonic vibration the crystal lattice is scarcely deformed at all and the microhardness increases about 60%. Apparently the decrease in grain size and increase in the packing density of the crystals increase the microhardness of the deposits produced in an ultrasonic field. Therefore, we cannot agree with the opinion of investigators who attribute the increase in microhardness of the deposits produced in an ultrasonic field to an increase in the number of foreign particles in the deposit resulting from their accelerated movement toward the electrode. Possibly, if one uses ultrasonic vibration of low intensity which does not produce cavitation and the variable sound pressure has a large effect, then no dispersion will occur and the cathode will be passivated as the result of the increasing number of foreign particles moving to the cathode. In this case the number of foreign particles included in the deposit will increase, and the microhardness will also increase. In our case ultrasonic vibration was accompanied by little cavitation and the resulting dispersion depassivated the cathode. The number of foreign particles included in the deposit decreased. In spite of this, the microhardness of the deposits increased because the deposit consisted of small crystals. Thus, our investigations showed that the microhardness of deposits produced in an ultrasonic field increases because the deposits consist of small crystals. The grain size and the packing density of the crystals are the determining factors in the change of the microhardness of the deposits.

The effect of ultrasonic vibration on the electrocrystallization of copper and silver was studied in [44] and [45]. The author found that deposits consisting of small crystals and having a high hardness and a fine structure can be produced at high current densities and that the parameters of the ultrasonic field have a great effect on crystal growth during electrodeposition. The size and shape of the crystals change when one changes the intensity and the frequency of ultrasonic vibration. Thus, ultrasonic vibration with a frequency of 330 kc leads to the formation of copper deposits with larger crystals.

Thus, the size of the crystals in electrolytic deposits depends on the parameters of the ultrasonic field. But, as a rule, when the ultrasonic intensity is no higher than $3 \cdot 10^4$ W/m^2 and the frequency is 20-23 kc the deposits consist of small crystals. We obtained interesting results which indicate the effect of ultrasonic vibration on the shape of the crystallites. Thus, in a study of the effect of ultrasonic vibration on ordinary crystallization from solutions, V. E. Kavalyunaite [46] noted that ultrasonic vibration has the greatest effect on the faces of crystals which are inclined at an angle to the front of the ultrasonic wave. This author also demonstrated experimentally the changes in the shape of crystals. A similar result was obtained in the case of electrodeposition of silver [44-46].

TABLE 11. Effect of an Ultrasonic Field on the Porosity of the Deposits

Nature of the deposit	Composition of the electrolyte, kg/m^3, and conditions of electrolysis	Thickness of the deposit, μ	Number of pores per cm^2	
			Without ultrasonic field	With ultrasonic field
Nickel	$NiSO_4 \cdot 7H_2O$, 240	3	130	58
	NaCl, 22	5	55	23
	H_3BO_3, 30	10	10	4
	pH = 5.53, t° = 50°C			
	δ_C = 150 A/m^2	15	3	0
Copper	$Na_4P_2O_7 \cdot 10H_2O$, 140	3	460	74
	$CuSO_4 \cdot 5H_2O$, 35	5	320	41
	HPO_4, 95	10	190	19
	t° = 40°C	15	120	4

The study of the effect of ultrasonic vibration on the texture showed that under certain conditions the [011] axis of the texture in copper deposits does not change. The study of the texture of nickel showed that ultrasonic vibration can either increase or decrease the degree of perfection of the texture, and this is apparently due to a change in ohmic polarization. It was also noted that the grain size in nickel deposits decreases. The deposits become uniform and bright. The external appearance of the cathode surface photographed with a microscope is shown in Figs. 21 and 22.

A microscopic study made by Ch. Kenakhen and D. Shlein showed that copper deposits produced in an ultrasonic field are more uniform and that the grain size decreases with increasing current density. X-ray analysis did not reveal any significant difference between the deposits obtained under ordinary conditions and in an ultrasonic field.

Weiner and Shiele obtained contradictory results on the effect of the intensity of ultrasonic vibration on the grain size. They came to the conclusion that the ultrasonic intensity plays an important role in electrodeposition of metals. Chromium, nickel, and iron deposits are uniform and consist of small grains when the ultrasonic intensity is lower than 1.8 W/cm^2. An increase in the ultrasonic intensity leads to an increase in grain size. Nickel deposits have a definite orientation with the axis parallel to the surface orientation of the cathode. Copper and silver form deposits with a rough structure when the intensity of ultrasonic vibration is low; an increase in the intensity of ultrasonic vibration results in deposits with smaller crystals. Weiner and Shiele also studied the effect of a pulsed ultrasonic field on the deposition of metals. The intensity of the pulses has an insignificant effect on electrocrystallization of nickel and chromium from acid electrolytes. Copper and silver deposits have different structures, depending on the intensity of the ultrasonic pulses. When the intensity of the ultrasonic pulses is low (about 0.5 W/cm^2) the copper and silver deposits consist of small crystals. With increasing pulse intensity the size of the crystals first increases and then decreases. The effect of the pulse frequency is different for different metals. Nickel deposits are uniform and consist of small crystals when the frequency of pulses is 1 pulse/sec. With increasing pulse frequency up to 20-50 pulses/sec the crystals become larger, and when the frequency reaches 100 pulses/sec the deposits consist again of small crystals. T. Kozan made a very detailed investigation of the structure of metals deposited in an ultrasonic field. He showed that ultrasonic mixing favors the formation of large grains in nickel deposits. The structure of the deposits was put in evidence by etching in a solution of HCl, H_2SO_4, and HNO_3. When the current density is high the grains have a tendency to decrease in size.

A. M. Ginberg showed that ultrasonic vibration can produce either small or large crystals in nickel deposits. This result was explained by the effect of ultrasonic vibration on the formation of nickel hydroxide in the cathode film. He also found that a considerable change in the grain size of the deposits occurs only at high current densities. If electrodeposition is carried out at current densities acceptable under ordinary conditions, ultrasonic vibration increases the grain size of the nickel deposits.

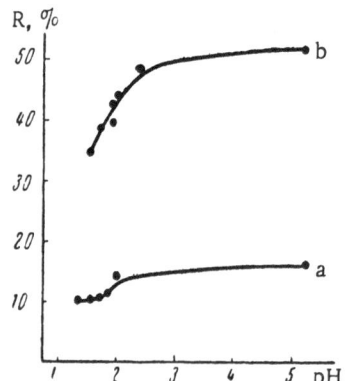

Fig. 23. Variation of the brightness of nickel deposits, R, %, expressed in arbitrary units in terms of the reflection coefficient of the surface. a) Under ordinary conditions; b) in ultrasonic field.

The porosity of nickel deposits produced in ultrasonic fields is much lower than the porosity of deposits produced under ordinary conditions, and therefore corrosion resistance increases (Table 11).

A decrease in the porosity of electrolytic deposits as the result of ultrasonic vibration was also noted by A. M. Lapides, N. S. Demchuk, V. I. Luk'yanov, A. M. Pavlov, and others. A. M. Lapides found that when the current density is 0.5-10 A/dm^2 copper platings up to 3 microns thick produced in ultrasonic field are practically free of pores, although ordinary platings of even greater thickness have many micro- and macropores. According to J. M. Odekerken, the effect of ultrasonic vibration on electrodeposition of metals consists essentially in a decreased porosity of the deposits. An interesting method of determining the porosity of nickel deposits was used by T. Kozan. A nickel film was deposited on a steel cathode and was used as a photonegative. The porosity was determined by the method of electroprinting. It was found that in all cases the porosity of deposits produced in an ultrasonic field is lower than that produced under ordinary conditions. Copper and nickel deposits 5-7 microns thick obtained by A. M. Ginberg, et al., were also less porous than deposits produced under ordinary conditions. But these authors found that the porosity of some platings increases when the ultrasonic intensity vibration is very high and cavitation very strong.

In the laboratory of the Kazan Chemicotechnological Institute it was found that the porosity of nickel deposits is decreased almost by a factor of $\frac{1}{2}$ when the deposits are produced in an ultrasonic field. When the current density was 30 mA/cm^2, the temperature was 30°C, and the thickness of the nickel deposit produced in an ultrasonic field was 20 microns the deposits had 12 pores per cm^2, while under ordinary conditions the number of pores was 23 per cm^2. The porosity of the deposits increases with increasing current density up to 40 mA/cm^2 and is 18 pores per cm^2 for deposits produced in an ultrasonic field and 24 pores per cm^2 for deposits produced under ordinary conditions.

To exhibit the pores in nickel plating deposited electrolytically on iron substrates we used a reagent consisting of 10 g/l of potassium ferricyanide adn 20 g/l of sodium chloride. The decrease in the porosity as the result of ultrasonic vibration is due essentially to a decrease in hydrogenation. Under ordinary conditions there is intense evolution of hydrogen at the cathode. This leads to the formation of craterlike depressions on the surfaces of the electrodes. Large pores reaching the substrate are formed at the bottoms of these craters. Hydrogen bubbles stick to the cathode and are retained on its surface for some time. The deposition of metal on these areas is rendered difficult, and pores are formed as the result. The ultrasonic field induces cavitation and mixing, so that hydrogen bubbles cannot stick to the cathode and screen it. All this decreases the number of micropores. In some cases there are no large pores at all in deposits produced in an ultrasonic field.

The number of micropores due to the structure of the electrolytic deposit also decreases, apparently due to the small crystals in the deposit produced in an ultrasonic field. Without the ultrasonic field the crystals are larger, hence the surface of the electrode is not completely covered with the metal; as a result, the number of pores increases. Also, the deposits produced in an ultasonic field are much denser, and this is also responsible for a smaller number of pores. The depassivation of cathodes, i.e., activation of cathode surfaces, also has a definite effect on the porosity of the deposits. Because of the depassivating action of ultrasonic irradiation, the number of particles adsorbed on the growing faces decreases, and therefore the number of pores in the deposit decreases because these particles prevent the growth of the faces and their coming together. The number of grooved pores also decreases because internal stresses are decreased; consequently, the probability of cracking of the deposit is much lower.

The same situation occurs in copper deposits. The porosity of copper deposits produced from acid electrolytes in an ultrasonic field is comparable to the porosity of plating obtained under ordinary conditions from

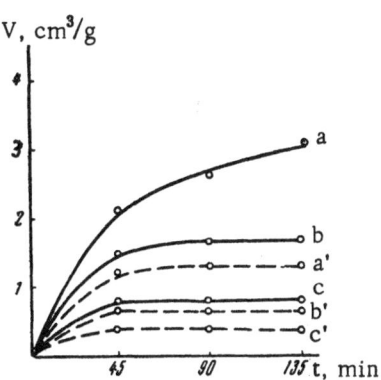

Fig. 24. Dependence of the amount of hydrogen evolved at different temperatures (V is the volume of hydrogen per gram of metal) as a function of sample heating time. a, a') Heating at 800°C; b, b') heating at 500°C; c, c') heating at 300°C; a, b, c) ordinary conditions; a', b', c') in ultrasonic field.

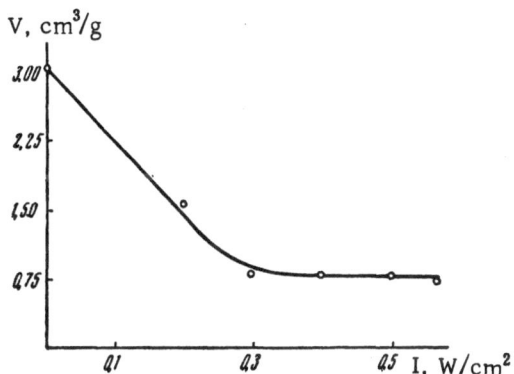

Fig. 25. Effect of the ultrasonic intensity field on the degree of hydrogenation of the deposit.

cyanides. The quality of the deposits obtained in an ultrasonic field is higher. Thus, under ordinary conditions the thickness of a corrosion-resistant deposit is 20 microns, while the thickness of corrosion resistant deposits produced in an ultrasonic field is 5 microns or a little thicker. The brightness of the deposits increases as the result of ultrasonic vibration. This is true in the case of nickel, cobalt, and iron. Deposits of other metals acquire a lighter and more uniform color under the influence of ultrasonic vibration.

The effect of ultrasonic vibration on the brightness of the deposit is shown in Fig. 23.

A. Roll explained the increase in brightness of the deposits under the action of ultrasonic vibration by the fact that electrocrystallization is somewhat different in an ultrasonic field. The discharging ions approach the cathode from a certain angle, which is determined by the velocity and amplitude of the ultrasonic vibrations. L. G. Plekhanov used the same theory to explain structural changes occurring in cathodic deposits under the influence of ultrasonic vibration. Müller and Kuss, M. I. Akinfiev, S. A. Shatsova, Yu. A. Fel'dman, N. S. Borodavko, and A. E. Ryabinina also found an increase in the brightness of deposits produced in ultrasonic field. These authors also found that the mechanical properties of brass deposits produced in an ultrasonic field improve. Bright brass plating is produced at current densities of 2-3 A/dm². Japanese scientists under the direction of Isid Osamutti investigated the principles of applying ultrasonic irradiation for bright nickel plating and found that deposits produced in an ultrasonic field are much brighter.

The effect of ultrasonic vibration on the brightness of deposits under different conditions of electrolysis (different current densities, different temperatures, and different pH of the electrolyte) was investigated in the laboratory of the Kazan Chemicotechnological Institute. The brightness was determined with a specially constructed apparatus, which measured the optical reflection coefficient of the cathode surface. The intensity of the reflected light was recorded with a photocell. Under ordinary conditions the deposits are matte and the deposits obtained at current densities of 10-20 mA/cm² have the maximum reflection coefficient. With increasing current density the brightness of ordinary deposits decreases and the reflection becomes less than 10%. The brightness of deposits produced in an ultrasonic field increases with increasing current densities up to very high values of the current density. Bright deposits are obtained over a much wider range of current densities. The brightness of nickel deposits decreases with decreasing pH of the electrolyte, which is due to the fact that the deposits obtained from acid nickel electrolytes are less smooth. The temperature has no significant effect on the brightness of the deposits. Deposits produced in an ultrasonic field have a brighter surface than the deposits produced under ordinary conditions, regardless of the temperature.

The analysis of the results obtained shows that the cause of the increase in the brightness of deposits produced in an ultrasonic field is neither the formation of a surface-active film on the cathode nor the directional

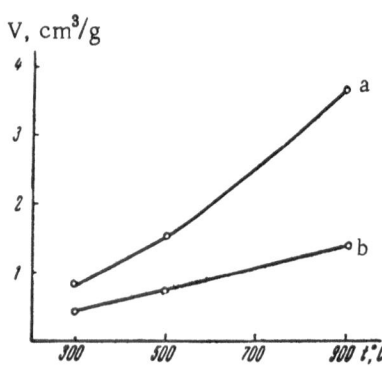

Fig. 26. Temperature dependence of
the amount of hydrogen evolved from
the nickel deposit. a) Under ordinary
conditions; b) in ultrasonic field.

orientation of crystallites, since the surface of the cathode is depassivated and the texture of the crystallites becomes less perfect.
Probably the change in the brightness of platings produced in an
ultrasonic field is due essentially to the change in the size of the
crystals and the lattice packing density.

In fact, electron microscopic analysis showed that the surface
of the brighter nickel plating consists of rather small crystallites
with a surface that appears to be smooth. Ordinary deposits have
an irregular large crystal structure.

There is practically no information on the internal stresses in
deposits produced in an ultrasonic field. T. Kozan found that with
a current density of 21.6 A/dm^2 the internal stresses in nickel deposits produced in an ultrasonic field are much larger than in ordinary deposits but decreases with increasing thickness of the deposit, although on the average the internal stresses in deposits produced in an ultrasonic field are always higher than in deposits produced without the field. In very thin plating produced in an ultrasonic field the stresses are low.

It is well known that the stresses depend on the amount of hydrogen included in the deposit. This is particularly true of metals in which discharge is accompanied by hydrogen liberation. However, there are almost
no data in the literature on the hydrogenation of deposits produced in an ultrasonic field. A. M. Ginberg and
A. P. Gorina indicate that ultrasonic vibration affects the hydrogenation of steel springs being plated with zinc,
and the degree of hydrogenation decreases only when the zinc plate has a thickness of 5-7 microns. When the
platings are thicker, the mechanical characteristics of the samples are lowered considerably, approaching those
of springs plated under ordinary conditions, i.e., without the ultrasonic field. S. Rich also found that zinc and
cadmium platings on steel springs in an ultrasonic field have no effect on the mechanical properties of the
springs. The mechanical properties were determined by bending tests. Thus, the bending angle before rupture
was 50°C before platings. The bending angle did not change after plating in an ultrasonic field. After plating under ordinary conditions the steel springs become more brittle and the critical bending angle is 20°C.

The effect of ultrasonic vibration on hydrogenation and internal stresses in electrolytic deposits of nickel
was studied in the laboratory of the Kazan Chemicotechnological Institute. The platings were subjected to
x-ray analysis and the effect of ultrasonic vibration on the texture was determined.

In all the experiments nickel was deposited from an electrolyte containing 280 kg/m^3 of NiSO$_4 \cdot$ 7H$_2$O.
The current density was varied from 200 to 700 A/m^2, the temperature of the electrolyte was 30°C, the pH was
approximately 2. The cathodes were prepared from copper foil degreased in acetone in a Soxhlet apparatus.
The size of the samples was varied, depending on the experimental conditions. For example, hydrogenation
was studied with samples 40 x 14 x 0.2 mm, while the internal stresses were measured on samples 60 x 5 x 0.2
mm. Electrolytic deposits were produced under ordinary conditions and in an ultrasonic field. The thickness
of the plating was 10-40 microns.

The ultrasonic field was generated with a UZM-1.5 generator, which provides a frequency of 20-23 kc
and an intensity of vibration in the electrolyte up to 10^4 W/m^2.

Determination of the Degree of Hydrogenation

During electrodeposition of nickel, metal ions and hydrogen ions are both discharged simultaneously at
the cathode. Some of the hydrogen atoms recombine and leave the surface of the cathode, while other atoms
are included in the deposit. Hydrogen is eliminated from the deposit by heating the samples in vacuum, using
the conventional method. The samples are heated in a quartz tube placed in a TK-30/200 furnace. Preliminary experiments showed that a plated sample placed in the reaction chamber in vacuum does not evolve any

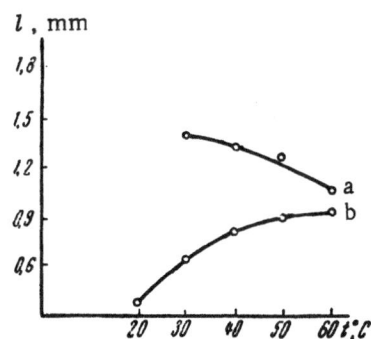

Fig. 27. Dependence of the stresses in the deposits on the temperature of the electrolyte. a) Under ordinary conditions; b) in ultrasonic field.

considerable amount of gas if it is not heated. The samples were heated at 300, 500, and 900°C for 135 min. The temperature was measured with a chromel-alumel thermocouple. After the experiment the system was cooled to room temperature and the pressure was measured with a VT-2 manometer.

Figure 24 shows the variation of the amount of hydrogen in deposits produced under ordinary conditions and in an ultrasonic field as a function of the heating time. The results show that the amount of hydrogen in deposits produced in an ultrasonic field is much lower than in deposits produced under ordinary conditions. The degree of hydrogenation decreases with increasing ultrasonic intensity (Fig. 25).

An analysis of the results for ordinary samples shows that the dependence of the amount of gas evolved on the evacuation time at elevated temperatures changes with the temperature. When samples are heated at 300 and 500°C the maximum amount of gas is evolved after 45-50 min of evacuation. Further increase in the heating time does not induce any further evolution of gases. At 900°C the gases are completely eliminated after 2-2.5 h.

The time after which the amount of gas evolved becomes constant is almost independent of the heating time for deposits produced in an ultrasonic field. This leads us to conclude that the strength of the metal-hydrogen bond in deposits produced in an ultrasonic field is of a different type. The variation of the amount of gas evolved with the heating temperature is shown in Fig. 26.

The amount of gas evolved increases almost fourfold when the temperature of the ordinary samples is increased from 300 to 900°C, while in deposits produced in an ultrasonic field this increase is insignificant. Thus, to remove the gases completely from deposits formed in an ultrasonic field it suffices to heat the samples to 300 or 500°C for 30 min. All these results indicate that the strength of the metal-hydrogen bond in deposits produced in an ultrasonic field is weaker than in deposits produced under ordinary conditions.

Study of Internal Stresses

The internal stresses in the deposits were studied by the flexible cathode method. The deviation of the lower end of the cathode was measured during the electrolysis with an MOV-1-1.5 micrometer. The cathode was a plate of copper foil $60 \times 5 \times 0.2$ mm. The cathode was rigidly secured in a holder. One side of the electrode was insulated with an organosilicon lacquer. The deposits were made at current densities of 100-700 A/m^2 at 300°C.

The investigation showed that nickel deposits produced in an ultrasonic field have very low internal stresses. The variation in internal stress with a change in the deposition conditions was very small. Thus, under ordinary conditions the internal stress in the deposits increased by $9996 \cdot 10^4$ N/m^2 when the current density was changed from low to high. In deposits produced in an ultrasonic field the change was only $4076 \cdot 10^4$ N/m^2.

The temperature dependence of the stress in deposits produced under ordinary conditions showed that an increase in the temperature from 30 to 60°C decreases the internal stresses (Fig. 27). When the temperature of the electrolyte is lower than 30°C the deposit cracks and the measured value of the stress is lower than the real value; consequently we did not measure the stress at these temperatures.

In an ultrasonic field the character of the variation in internal stress with temperature is different. When the temperature is increased from 20 to 60°C the stress in the deposit increases from $3822 \cdot 10^4$ N/m^2 to $10,829 \cdot 10^4$ N/m^2. The magnitude of the stress does not change with increasing temperature.

This anomalous dependence of stress on the temperature of the electrolyte is probably due to a decrease in the effect of the ultrasonic field with increasing electrolyte temperature.

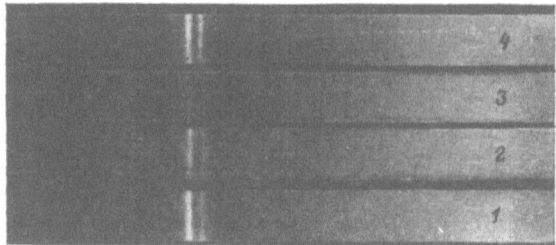

Fig. 28. X-ray diagrams of electrolytically deposited nickel.
1, 2) In an ultrasonic field; 3, 4) under ordinary conditions; 2, 3)
before annealing; 1, 4) after annealing.

A comparison of the data on internal stresses and the adhesion of the deposit to the substrate showed that deposits produced in an ultrasonic field have lower stresses and the adhesion of the deposit to the substrate is better. In an electrolyte with a pH of 5-6 and a temperature of 40°C, not very bright uniform deposits with internal stresses of $3822 \cdot 10^4$ N/m^2 were produced in an ultrasonic field. Under ordinary conditions the deposits cracked and split away from the substrate. Thus, the decrease in the amount of hydrogen occluded in the deposits in an ultrasonic field decreases the stresses and improves the ductility and elastic properties of the deposit.

It is well known that hydrogen can penetrate into electrolytic deposits in different ways. During the growth of the deposit atomic hydrogen is adsorbed on the surface of the electrode and forms chemical compounds and also penetrates into the deposit as the result of the capture of gas bubbles by the growing faces of the deposit. The cavitational effect of ultrasonic vibrations "sucks" molecular hydrogen away from nickel. The ultrasonic field decreases the amount of hydrogen captured mechanically because gas bubbles formed on the surface of the cathode are removed from the electrolyte by the ultrasonic vibration. Apparently, ultrasonic vibration affects the catalytic activity of the surface of the cathode, accelerates the recombination of hydrogen atoms into molecules, and, as a result, decreases the amount of atomic hydrogen in the crystal lattice.

X-Ray Analysis

It is well known that internal stresses occurring in metals change the lattice constants. This change can be seen in the x-ray diffraction diagrams of the electrolytic deposits. One can see that the deposits produced in an ultrasonic field have lower internal stresses and the x-ray diagrams are different from those for deposits produced under ordinary conditions.

We used Fe, α, β-radiation and the RKÉ camera, which makes it possible to obtain five photographs on one film. The sample was photographed at an angle of 72°. The x-ray diagrams are shown in Fig. 28. It can be seen that the diffraction diagrams are different for deposits produced under different conditions. Photographs 1 and 2 refer to deposits produced in an ultrasonic field and photographs 3 and 4 to deposits produced under ordinary conditions.

Comparison of photographs 2 and 3 shows that the interference fringes corresponding to deposits produced under ordinary conditions are widened. Apparently, this widening is due to internal stresses. The fringes relative to deposits produced in an ultrasonic field are sharp. (The photographs were made immediately after removal of the samples from the electrolyte.)

The samples were then heated at 900°C in vacuum and analyzed again. The results are shown in Fig. 28 (1 and 4). The x-ray diagrams of samples produced in an ultrasonic field are not significantly affected by the extraction of hydrogen in vacuum: the interference fringes remain sharp. Photographs 3 and 4 show the difference between the deposits produced under ordinary conditions before and after heating.

Fig. 29. X-ray photographs of electrolytic deposits of nickel. a) Under
ordinary conditions; b) in ultrasonic field.

Investigation of the Texture of Deposits

The study of the effect of ultrasonic vibration on the texture of nickel deposits showed that the degree of perfection of the orientation of crystallites depends on the deposition conditions. When the current density is 100 A/m^2 the texture of the deposits produced in an ultrasonic field is more perfect than that of deposits produced under ordinary conditions. When the current density is 300 A/m^2 the texture of the deposits produced under ordinary conditions and in an ultrasonic field is the same. When the current density is 400-500 A/m^2 the orientation disappears almost completely in the deposits produced in an ultrasonic field, while the degree of perfection increases in deposits produced under ordinary conditions.

The study of the combined effect of ultrasonic vibration and temperature of the electrolyte on the perfection of orientation showed that the degree of perfection of deposits produced in an ultrasonic field is lower than that of deposits produced under ordinary conditions.

When the current density is 200 A/m^2 and the temperature is below 30°C the deposits produced in an ultrasonic field are not clearly oriented. Deposits produced under the same conditions but without the ultrasonic field have no texture. The x-ray diagrams of the deposits are shown in Fig. 29.

The decrease in the number of uniformly oriented crystallites as the result of the ultrasonic field is apparently due to a decrease in the degree of hydrogenation and a decrease in polarization.

W. Wolfe and other investigators who have studied the electrodeposition of copper in ultrasonic fields showed that ultrasonic vibration changes the preferential orientation of crystallites in copper deposits. The effect of ultrasonic vibration was similar to the effect of ordinary mechanical mixing. Apparently, ultrasonic vibrations affect the texture only in dilute solutions, where the concentration polarization is more intense (Table 12).

Mechanism of the Effect of Ultrasonic Vibration on the Physicochemical Properties of Electrolytic Deposits

Ultrasonic vibration affects the electrocrystallization process and changes the structure of the deposited metal.

These changes show that ultrasonic vibration has a direct effect on all stages of polarization during electrocrystallization of nickel and copper and also on ohmic polarization. The physical state of the surface of the cathode is altered by ultrasonic vibration. Structural changes occurring on the surface of the cathode were recorded with an electron microscope. The photographs show that the surface of nickel deposits produced in an ultrasonic field consist of rather small crystals with a "smooth" surface united into aggregates of equal size. On the surface of deposits produced under ordinary conditions there are large crystal blocks separated by irregularly distributed small-grained aggregates.

The changes in surface structure under the influence of acoustic vibration is the result of changes in the growth process of the deposits in connection with the change in crystal growth rate and rate of foundation of new crystallization centers. The hydrodynamic conditions in the electrode layer apparently play the principal

TABLE 12. Measurements of the Relative Orientation of Copper Deposits on the Five-Point Scale

Conditions	Concentration of solution, kmole/m^3		Current density, A/m$^2 \cdot 10^{-1}$	Relative orientation
	CuSO$_4$	H$_2$SO$_4$		
Not mixed	0.17	0.02	8	2
Mixed	0.17	0.02	8	4
In ultrasonic field	0.17	0.02	8	4
Not mixed	0.17	0.02	20	1
Mixed	0.17	0.02	20	4
In ultrasonic field	0.17	0.02	20	4
Not mixed	0.86	0.02	20	5
Mixed	0.86	0.02	20	5
In ultrasonic field	0.86	0.02	20	5

role in this process. The hydrodynamic conditions most probably determine the specifics of the mechanism of the effect of ultrasonic vibrations on the growth of the crystal lattice, since the formation of deposits consisting of small crystals under the influence of ultrasonic vibration is difficult to explain from the viewpoint of classical electrochemistry. According to the familiar theory of the formation of new metal nuclei during electrocrystallization, the decrease in polarization and the decrease in the passivity of the surface of the cathode should lead to the formation of large crystals. In an ultrasonic field this is not always so. Thus, the separation potential of nickel is shifted 200 mV toward the equilibrium potential. A comparison of the variation of potentials with time shows that ultrasonic vibration decreases the passivation rate of the cathode.

Structural changes occurring in the deposits during electrolysis in an ultrasonic field, as well as degassing and adsorption induced by ultrasonic vibration, decrease the degree of hydrogenation of nickel deposits.

According to the results of our investigations, the amount of hydrogen included in deposits produced in an ultrasonic field is half that in deposits produced under ordinary conditions. This decrease is due, first of all, to the capture of molecular hydrogen by the growing faces of the deposit and also to a decrease in the adsorption of atomic hydrogen on the surface of the cathode. Analysis of the rate at which hydrogen is evacuated from the deposits showed that the kinetics of the evolution of gases from metals deposited with and without ultrasonic vibration is different. In the case of metals deposited with ultrasonic vibration the main volume of hydrogen is evolved at 500°C. It is necessary to heat the sample for 30 min to eliminate all the hydrogen. In the case of metals deposited under ordinary conditions the complete elimination of hydrogen requires higher temperatures and a longer time. All these results indicate that the nature of hydrogen trapped in nickel deposited in an ultrasonic field is different from that trapped in nickel deposited under ordinary conditions and that the nickel-hydrogen bond is much weaker in nickel deposited in an ultrasonic field.

The weakening of the nickel-hydrogen bond in deposits produced in an ultrasonic field can be explained by the change of the force field in the crystal lattice of nickel deposited in an ultrasonic field. An electron microscopic study showed that the structure of nickel on the surface of the deposit is very different. X-ray analysis showed that the crystal lattice of the metal deposited in an ultrasonic field is much less deformed than the crystal lattice of the metal deposited under ordinary conditions and is much closer to that of the equilibrium state. But if the lattice constants are changed as the result of ultrasonic vibration, the force field of the lattice must also change, and therefore the solubility of hydrogen in the metal is also modified. According to N. A. Galaktionova, the strength of the bond between hydrogen and the metal depends on the depth of penetration of the hydrogen proton into the electron shells of the metal atoms. The more intense the force field of the crystal lattice the deeper the penetration of the proton into the electron cloud and the stronger the metal-hydrogen bond. The degree to which the metal is hydrogenated and the strength of the metal-hydrogen bond also depend on the physical state of the cathode surface. This dependence is readily observed by examining the effect of ultrasonic fields of different intensities on hydrogenation. As the ultrasonic field intensity is lowered, the degree to which the deposited metal is hydrogenated gradually decreases, then stops, and an increase in the intensity of the field at this point does not lower the degree of hydrogenation. This dependence of the degree of hydrogenation on the intensity of the ultrasonic field can probably be explained by the changes in the surface state and the structural change in the bulk of the deposit resulting from ultrasonic vibration. In fact, the structure of the metal changes gradually under the influence of ultrasonic vibrations of low intensity. Small as well as large crystals of metal are deposited. These crystals are stabilized on the surface and within the bulk of the deposit as the result of ultrasonic vibration, and further increase in the ultrasonic intensity no longer has any effect on the occlusion capacity of the nickel deposit.

The change of the physical state of the surface of the electrode under the influence of ultrasonic vibration is confirmed in the literature. Thus, in [15] and [31] it was shown that ultrasound changes the value of the constant << a >> in the Tafel equation:

$$\eta = a + b \log I,$$

where η is the overvoltage, I is the current, and b is a constant.

The value of the constant << a >> depends on the nature of the cathode. This value is a minimum when the interatomic spacing in the crystal lattice is close to 2.75 A. As the interatomic spacing in the crystal lattice approaches the value of 2.75 Å during electrolysis in an ultrasonic field the value of << a >> decreases and the overvoltage decreases. However, if under the influence of the ultrasonic field the crystal lattice is changed in such a way that the interatomic spacing departs further and further from 2.75 Å, then the value of << a >> increases and the overvoltage increases. Possibly this difference is the reason for the contradictory results published with regard to the effect of ultrasonic vibrations on the electrochemical deposition of hydrogen and metals.

The structural changes occurring in the metal deposited in an ultrasonic field change the physicochemical properties of the deposits. Thus, internal stresses in deposits produced in an ultrasonic field are 2-3 times lower than in deposits produced under ordinary conditions. Under ordinary conditions of electrolysis the internal stresses in the deposits depend greatly on the conditions of electrolysis. For example, an increase in the current density from 100 to 700 A/m^2 increases the internal stresses almost fourfold. When electrolysis is carried out in an ultrasonic field the overvoltage changes by only a factor of 1.5 under the same conditions. The internal stresses in nickel deposits produced in an ultrasonic field depend very little on the acidity of the electrolyte or the thickness of the deposit. This is due to the stabilization of the electrodeposition process by ultrasonic vibration: the alkalinity of the cathode layer is eliminated, the acidity of the electrolyte does not change, the surface of the cathode is depassivated, etc. As the result, the structure of nickel layers deposited in an ultrasonic field is the same at any given time of electrodeposition. Under ordinary conditions the structure of the deposit is nonuniform throughout the thickness of the deposit.

The structure of the deposits plays a decisive role in the type of internal stress. Deposits with deformed crystal lattices are characterized by the highest internal stresses. Deformations of the lattice may be induced by inclusions of impurities (including hydrogen) and also high overvoltage resulting from the discharge of metal ions on the cathode. The amount of hydrogen occluded in nickel is less in deposits made in an ultrasonic field, and therefore the potential of the deposition of nickel is lower. Consequently, the potential gradient in the electric double layer decreases and the average vibrational energy of the discharging ions decreases because the ions enter the lattice with a lower reserve of energy and the constants of the crystal lattice of the deposit being formed are closer to the equilibrium values, i.e., the lattice is less deformed.

The grain size, which determines the degree of perfection of the texture, changes in deposits produced in an ultrasonic field. The lower degree of deformation of the crystal lattice and the lower stresses in the deposit are apparently also reasons for less texturization of the deposits produced in an ultrasonic field. The decrease in the degree of orientation may also be induced by the decreased separation potential of nickel. Therefore, one must take into account all these factors in determining the mechanism of the orientation of crystals in an ultrasonic field. Ultrasonic vibration changes the degree of perfection of the texture but has practically no effect on the direction of the axis of orientation. Most of the crystals are oriented along the [001] axis either in the presence or absence of ultrasonic vibration. At temperatures between 30 and 50°C the main direction of orientation is along the [011] axis.

Since the grain size changes, such an important factor as texture does not affect the physicomechanical properties of nickel deposits, particularly the microhardness and brightness. In spite of a decrease in the degree of orientation of crystallites, nickel platings produced in an ultrasonic field are harder and more brilliant than those produced under ordinary conditions.

The enhanced brightness of deposits produced in an ultrasonic field is due essentially to a change in size of the crystallites and the packing density. These factors are apparently also the cause of the higher microhardness of deposits produced in an ultrasonic field. In fact, the decrease in the degree of hydrogenation and the decrease in the deformation of the crystal lattice should decrease the hardness of the deposit, but this is not the case. Apparently, the decrease in the grain size and the increase in packing density have a greater effect on the microhardness, and the role of structure turns out to be so significant that it prevails over all the other factors. As a result, the microhardness increases. The results of our investigations lead us to conclude that the microhardness of deposits produced in an ultrasonic field increases if the deposits consist of small crystals. In this case the decrease in the amount of hydrogen and other foreign elements in the deposit does not decrease their microhardness.

Effect of Ultrasound on the Uniformity of the Distribution Electrodeposition Metal on the Cathode Surface

The data on the effect of ultrasonic vibration on the mixing of the electrolytic bath are contradictory. Some authors [32] consider that the metal yield from the electrolyte decreases with the depth of the electrolyte because of decreasing polarization. However, direct measurements of the distribution of metal on the cathode surface show that this is not so. L. Bergmann [15] indicates that electrolytically deposited copper, nickel, cadmium, zinc, silver, and gold are more uniform throughout the thickness when deposited with ultrasonic treatment.

The authors in [11] studied the effect of ultrasonic vibration on the electrocrystallization of metals and found that the throwing power of the bath improves in the case of electrodeposition of nickel, copper, and zinc. In [37] the same results were obtained in the case of electrolytic deposition of zinc, cadmium, copper, nickel, and tin. In [17] it was also found that the distribution of the metal over the surface of the cathode improves under the influence of ultrasonic vibration in the case of electrodeposition of nickel, copper, silver, and chromium. A number of investigators have found that ultrasonic vibration is particularly useful in the plating of machine parts with intricate shapes. The effect of ultrasonic vibration on the throwing power of the bath during deposition of copper was investigated in [32] and [33]. The author concluded that the throwing power of the bath is diminished as the result of the depolarizing action of ultrasound and that the distribution of the metal over the surface of the cathode becomes less regular. However, in his later investigation concerning the distribution of metals on cathodes with complex shapes the same author showed that the distribution of the metal deposited from copper pyrophosphate becomes more uniform under the influence of ultrasonic vibration, while ultrasonic vibration has no significant effect on the uniformity of plating deposited from nickel sulfate [33]. The author explained these results by the fact that during deposition of copper from pyrophosphate the polarization of the cathode is essentially concentration polarization, and ultrasonic vibration therefore renders the supply of the electrolyte to all the areas of the cathode more uniform, so that the current distribution on the cathode also becomes more uniform. During electrolysis in nickel sulfate the polarization of the cathode is essentially chemical and the decrease in polarization and change in the yield under the influence of ultrasonic vibration are small, and therefore the effect of ultrasonic vibration on the uniformity of the plating is insignificant.

We used the radiographic method [47] to determine in detail the distribution of metal on the cathode and the uniformity of dissolution of the anode under the influence of the ultrasonic field.

The experimental conditions were as follows: 200 kg/m^3 of $CuSO_4 \cdot 5H_2O$; 60 kg/m^3 of H_2SO_4; $\delta_c = 300$ A/m^2; t = 30°C; $f = 21$ kHz; I = $3 \cdot 10^4$ W/m^2.

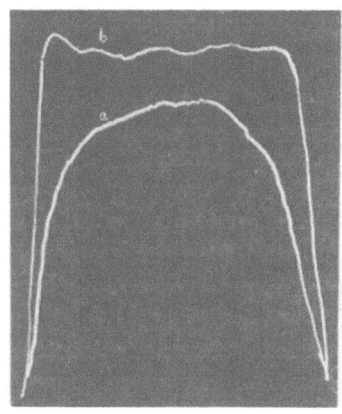

Fig. 30. Autoradiograph of copper deposits. a) Distribution of the metal in a deposit produced under ordinary conditions; b) distribution of the metal in a deposit produced in an ultrasonic field.

Copper was deposited electrolytically on a cathode coated with a thin film of radioactive thallium. The thickness of the deposited plating was 20 microns after 20 min of electrolysis. Then autoradiographs were made by a method developed in our laboratory. Analysis of the autoradiograph (Fig. 30) shows clearly that under the experimental conditions indicated the electrolytic deposition of copper in an ultrasonic field (b) is much more uniform over the surface of the cathode than under ordinary conditions (a) (the width of the sample is along the x axis and the thickness of the plate along the y axis). Similar results were obtained in the case of anodic dissolution of copper. To determine the effect of the ultrasonic field on the distribution of the metal it is apparently necessary to take into account the more regular polarization of the cathode surface. Electrocrystallization of metals on the cathode is very sensitive to the ultrasonic frequency. When the frequency is 18-30 kc the thickness of the deposit is essentially uniform. When the frequency is 50-100 kc the surface of the electrode develops ripples, as observed by W. T. Young and H. Kersten [60]. The troughs and the crests alternate at distances equal to half the wavelength. Figure 31 is a photograph of such a surface. Some authors explain this phenomenon by assuming that the change in polarization is particularly large at the sections of the cathode corresponding to the ultrasonic wave crests. J. Guitton ascribed the ribs of the electrode to energetic oxidation of the metal surface due to cavitation.

Role of Cavitation and Mixing in Electrodeposition of Metals

Studies of the effect of cavitation on the electrodeposition of metals have shown that cavitation increases with increasing sound intensity. The dependence of the cavitation threshold on the ultrasonic frequency and intensity is shown in Table 13.

At the electrolyte-electrode boundary cavitation handicaps electrodeposition with increasing ultrasonic intensity. Using the concepts advanced in [48], one can represent a cavitation void at the surface of the metal by the contact diagram in Fig. 32.

This diagram indicates that the greater the number of such voids, the more difficult electrocrystallization becomes. The force of the imploding cavities breaks up the surface of the electrode. The most advantageous ultrasonic intensity in this case is below $30 \cdot 10^4$ W/m². It should be noted that when cavitation is significant the electrolyte may become locally overheated, particularly at the phase separation boundary, where an effect similar to boiling may occur [19]. Trillat [19] subjected aluminum to anodic oxidation in an ultrasonic field of 900-1000 kc and found that the growing anodic films are deformed mechanically and have visible discon-

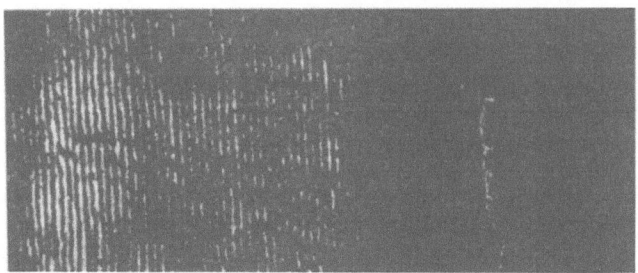

Fig. 31. View of the surface of the electrode resulting from deposition in an ultrasonic field with a frequency of 50-100 kHz.

TABLE 13

Frequency, kc	Required for initiation of cavitation:	
	Sound pressure, $s/m^2 \cdot 10^{-4}$	Intensity, $W/m^3 \cdot 10^{-4}$
15	4.9-19.6	2.6
175	39.2	10
365	68.6-196	33-270
500	117.6-245	100-400
3300	2254-2940	35000-60000

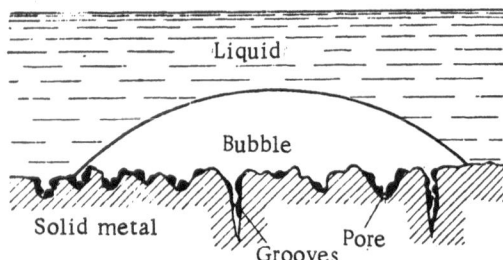

Fig. 32. Diagram of a cavitation bubble on the surface of the metal.

tinuities. In some places the films were detached from the base metal. Good quality films are obtained when the frequency of ultrasonic vibration is 20-30 kc.

Let us investigate the effect of mixing on different types of polarization.

According to the theoretical calculations by S. V. Gorbachev, L. P. Kholpanov, and I. I. Aryamova, the current density inducing chemical and concentration polarizations during electrolysis depends on the activation energy of the system, the activity of the substance, the diffusion coefficient, and the rate of rotation of the disc electrode:

$$I = zF\left[\frac{\kappa_1 a e^{-\frac{E}{RT}}(D + \kappa_B \sqrt{\omega})}{D + \kappa_B \sqrt{\omega} + \kappa_1 U_B e^{-\frac{E}{RT}}}\right]$$

where I is the current, E is the activation energy, ω is the rate of rotation of the disc electrode, D is the diffusion coefficient, a is the activity of the reacting substance, k_1, k_B, and U_B are constants, F is the Faraday number, and z is the ionic charge.

If we assume that ultrasonic vibration has only a mixing effect, this equation can then be used to characterize the effect of the ultrasonic field on the concentration and chemical polarizations. In reactions in which the effective activation energy is low, i.e., the discharge process is rapid, chemical polarization does not slow down the cathodic reaction, and concentration polarization plays the main role. In this case the term $e^{-E/RT}$ is large and

$$\kappa_1 U_B e^{-\frac{E}{RT}} \gg D + \kappa_B \sqrt{\omega},$$

hence we cannot neglect the term $D + k_B\sqrt{\omega}$, and the equation for the process determined by concentration polarization takes the form

$$I = \frac{zFa}{U_B}(D + \kappa_B \sqrt{\omega}).$$

The current density increases with the rate of rotation of the disc. In our case we may assume that ω is approximately equal to the ultrasonic frequency, i.e., when concentration polarization plays the determining role in electrolysis, ultrasonic vibration directly affects the increase of the discharge current, and this current depends on the ultrasonic frequency. However, if the activation energy is high, chemical polarization plays the determining role, and then

$$D + \kappa_\text{в} \sqrt{\omega} \gg \kappa_1 U_\text{в} e^{-\frac{E}{RT}}.$$

Neglecting the term $k_1 U_B e^{-E/RT}$ in the general equation for the current density, we obtain

$$I = zF\kappa_1 a\, e^{-\frac{E}{RT}}.$$

It can be seen that the current density becomes independent of the frequency of ultrasonic vibration. From this we might conclude that ultrasonic vibration, which has only a mixing effect, does not affect the chemical polarization during electrocrystallization. However, this is not quite so. We assumed earlier that ultrasonic vibration has only a mixing effect. If, however, we take into account the different cavitation phenomena which occur in intense ultrasonic fields we may assume that ultrasonic vibration will also have an effect on the electrochemical reaction proper. Consequently, when the electrode reaction is limited by concentration polarization the reaction will be accelerated by an ultrasonic field with low intensity, which has essentially a mixing effect. If one wishes to accelerate the reaction by decreasing chemical polarization, then one should use ultrasonic vibration of high intensity, which induces cavitation phenomena, since the mixing effect of ultrasonic fields not inducing cavitation has no effect on chemical polarization.

In studying the effect of ultrasonic vibration on ohmic polarization one must take into account the fact that an ultrasonic field of high intensity leads to a strong dispersion effect, which may destroy the metal deposit.

It should be of interest to investigate the direct effect of ultrasonic vibration on all types of polarization by depositing metals under different types of polarization according to the method of S. V. Gorbachev. Also, as we noted earlier, the study of electrode processes in ultrasonic fields requires that one take into account hydrodynamic effects in the electrode layer and consider the nucleation and the growth of the crystal phase from this viewpoint.

Effect of an Ultrasonic Field on the Metal Yield

Different investigators have found different effects of ultrasonic vibration on the metal yield.

Some investigators have found that ultrasonic vibration increases the yield of nickel and copper, while others have found that it decreases the yield. In [36-37] it was found that the yield of nickel is increased 90% under industrial conditions.

A number of authors showed that the yield of nickel decreases considerably under the influence of ultrasonic vibration and that it is greatly dependent on the intensity of the ultrasonic vibration. Probably this decrease in the yield of nickel is due to the fact that ultrasonic vibration has different effects on the polarization during deposition of nickel and evolution of hydrogen. Polarization resulting from the evolution of hydrogen decreases under the influence of ultrasonic vibration; therefore, when the pH of the solution and the current density are low the current is utilized mainly in the liberation of hydrogen, and the yield of nickel drops. When the current density is high and the acidity of the electrolyte is low the nickel yield may be increased by the ultrasonic field.

A detailed investigation of the effect of ultrasonic vibration on electrodeposition of chromium was made in [26]. The authors concluded that ultrasonic vibration with a frequency between 17 and 80 kc does not affect the dependence of the yield on the current density.

In such studies it is necessary to remember that ultrasonic vibration of different frequencies and different intensities has different specific effects on the cathode process and, as a result, the yield may be affected differently by different ultrasonic vibrations.

Many authors have found that the yield of the metal increases when the intensity of ultrasonic vibration with a frequency between 18 and 25 kc is lower than $3 \cdot 10^4$ W/m^2. Further increase in the frequency, and particularly of the intensity, of vibration results in cavitation, and when the intensity is $5 \cdot 10^4$ to $10 \cdot 10^4$ W/m^2 the surface of the electrode begins to break up.

Spongy and dendritic deposits break up very easily. This effect is particularly clear in the case of tin, lead, and other metal deposits [11].

Some of the Technological Conditions of Electrodeposition of Metals in an Ultrasonic Field

In what follows we describe the compositions of electrolytes and the conditions of deposition of metals in an ultrasonic field and also the characteristics of electrodeposition processes and the quality of the deposits. The sequence of operations is the same in an ultrasonic field as under ordinary industrial conditions of deposition. First of all, the surfaces are thoroughly cleaned, washed, and degreased. Ultrasonic treatment is strongly recommended for these operations. When machine parts are relatively clean, metals can be deposited without subjecting the parts to special chemical treatment. In this case the parts are immersed directly in the electrolyte and subjected to ultrasonic vibration for 10-15 min before the current is turned on.

Zinc

Zinc coatings of very good quality can be obtained at very high current density in an ultrasonic field. Yu. M. Bystrov and N. A. Evdokimov deposited zinc from sulfate solutions at a current density of $2 \cdot 10^3$ A/m^2 in an ultrasonic field; without the ultrasonic field the acceptable current density is 10^2 to $2 \cdot 10^2$ A/m^2. V. A. Mikhailov deposited zinc at current densities from $5 \cdot 10^2$ to $15 \cdot 10^2$ A/m^2 from zincates. The zinc deposits are less porous and brighter than those obtained under ordinary conditions. According to N. T. Kudryavtsev and A. M. Smirnova, the yield in an ultrasonic field increases by a factor of 3-5 in the case of cyanides and acid electrolytes and by a factor of 8 in the case of zincates. The microhardness of the deposits does not change. Ultrasonic vibrations of high frequency (of the order of 1200 kc) and high intensity lead to the formation of a rippled surface.

V. I. Falicheva, P. A. Margolin, and I. P. Mochalina studied the formation of zinc coatings deposited from cyanide solutions in an ultrasonic field under industrial conditions. These authors also carried out laboratory experiments. They used a 240-liter bath for semi-industrial tests. The PM-1 and 5D-1 vibrators were mounted in the bottom of the bath and several vibrators were suspended at different points in the bath. They plated flat pieces and various machine parts with zinc (clamps, nuts, bolts, sockets, etc.). The limiting current density for electrolytes rich in zinc was raised considerably by ultrasonic vibration. When the concentration of zinc is lower than 40 g/liter the deposits are of poor quality and the limiting current density can be increased very little in the presence of the ultrasonic field. The authors found that when the specific volume intensity of the ultrasonic field is 6-8 W/liter of solution the cathode current density can be increased to 11 A/dm^2. Under these conditions the rate of deposition reaches 1.2-2 microns/sec. The distribution of the metal remained satisfactory when larger equipment was used. It was found that the effectiveness of ultrasonic vibration depends greatly on the positions of the vibrators and of the parts to be plated in the electrolytic bath [68].

B. Brown made a semi-industrial investigation and found that coating of cylinders with zinc deposited from cyanide solutions is accelerated in an ultrasonic field. The ultrasonic vibration is effective even when the articles to be plated are not subjected to preliminary treatment.

The basic information concerning the conditions of deposition of zinc in an ultrasonic field are given in Table 14.

TABLE 14. Technological Conditions for Electrodeposition of Zinc

Composition of the electrolyte, kg/m^3	Conditions of electrolysis		Ultrasonic conditions		Quality of the deposit	Source of data
	Current density, $A/m^2 \cdot 10^{-3}$	Temp., °C	Frequency, kc	Intensity, $W/m^2 \cdot 10^{-4}$		
(ZnSO₄), 250 (NH₄Cl), 15 (CH₃COONa), 15	0.4	47	1200	250	Ribbed deposits	[60]
(Zn(CN)₂), 30 (NaCN), 30 (NaOH), 25	0.4	47	1200	250	Deposits not bright	[60]
(ZnSO₄), 215 (Al₂(SO₄)₃), 20 (Na₂SO₄), 100	2.0	20	28	0.8	Less porous deposits	[29]
(ZnO), 50 (NaOH), 120 (NaCN), 120	0.5–1.5	50	20	1	Good quality deposits	[37]
(NaCN), 80–90 (NaOH), 70–80 (Zn), 28–30		75	21	–	Quality improved	[35]
(ZnSO₄), 250 (Al₂(SO₄)₃), 40 (Na₂SO₄), 1 Dextrin, 10	2.0	16	2.5	–	Ductility of deposit improved	[21]

TABLE 15. Technological Conditions for Electrodeposition of Chromium

Composition of the electrolyte, kg/m³	Conditions of electrolysis		Ultrasonic conditions		Quality of the deposit	Source of data
	Current density, A/m²·10⁻³	Temp., °C	Frequency, kc	Intensity, W/m²·10⁻⁴		
Cyanide electrolyte	–	–	20	–	Quality of deposits improved	[59]
(CrO₃), 100–450 (H₂SO₄), 10	1.5–4.0 —	20–60 —	17 80	— —	Deposits comparable to those obtained under ordinary conditions	[26]
	22.0	–	20 40	2.5	Throwing power improved	[61]
	–	–	20	–	Effect same as that of ordinary mixing	[17]
(CrO₃), 250 (H₂SO₄), 1.5	1.0–1.5	40	21	5.8	Quality of the deposits not improved; turbidity of the electrolyte increased	[11]
(CrO₃), 250 H₂SO₄, 2.5	Higher than 1.0	45	22 300, 600	–		[28]

TABLE 16. Technological Conditions for Electrodeposition of Cadmium

Composition of the electrolyte, kg/m³	Conditions of electrolysis		Ultrasonic conditions		Quality of the deposit	Source of data
	Current density, A/m²·10⁻³	Temp., °C	Frequency, kc	Intensity, W/m²·10⁻⁴		
(CdO), 30–50 $(NaCN)$, 80–130 (Na_2SO_4), 30–60 $(NiSO_4)$, 0.8–2	1.8	–	20	1	Deposit consists of small crystals	[37]
$(NaCN)$, 75 (CdO), 25 $(NaOH)$, 15	0.5	47	1200	250	Bright deposits	[60]
Cyanide electrolyte	0.8	–	20	–	Uniform, bright deposit	[59]

TABLE 17. Technological Conditions for Electrodeposition of Iron and Cobalt

Composition of the electrolye, kg/m³	Conditions of electrolysis		Ultrasonic conditions		Quality of the deposit	Source of data
	Current density, A/m²·10⁻³	Temp., °C	Frequency, kc	Intensity, W/m²·10⁻⁴		
$(FeCl_2)$, 300 $(CaCl_2)$, 150	0.5	47	1200	–	Improves	[60]
$(FeSO_4)$, 200 $(MgSO_4)$, 100	0.5	47	1200	–	Improves	[60]
$(CoSO_4)$, 280 (KCl), 15 (Na_2SO_4), 50 (H_3BO_3), 30 pH = 4, 5, 6	0.5	20	21	5.8	Polarization decreases	[11]
$(Co(NH_3)_4SO_4)$, 200	0.5	47	1200	500	Rippled deposits	[60]

TABLE 18. Technological Conditions for Electrodeposition of Nickel

Composition of the electrolyte, kg/m^3	Conditions of electrolysis		Ultrasonic conditions		Quality of the deposit	Source of data
	Current density, A/m$^2 \cdot 10^{-3}$	Temp., °C	Frequency, kc	Intensity W/m$^2 \cdot 10^{-4}$		
(NiSO$_4$(NH$_4$)SO$_4$), 50 (NH$_4$CN), 50 (ZnSO$_4$), 10	0.4	47	1200	500	Rippled deposits	[60]
(NiSO$_4$), 40 C$_3$H$_4$(OH)(COONa)$_3$, 35	0.4	20	34	0.3	Bright deposits	[21]
Sulfate-chloride electrolyte containing boric acid	3.0	40	16	—	Brightness, hardness, and texture improved	[38]
(NiSO$_4$), 280 pH=4.5	0.5	20	21	5.8	Separation potential decreased	[11]
(NiSO$_4$), 280 (KCl), 15 (Na$_2$SO$_4$), 65 (H$_3$BO$_3$), 30 pH=5						
(NiSO$_4$), 170 (H$_3$BO$_3$), 26 pH=5.4	1.6	—	28	0.8	Less porous deposits	[29]
Acid electrolyte	0.8	20	20	1	Bright deposits	[37]
(NiSO$_4$), 300 (H$_3$BO$_3$), 35–45 (NaCl), 25–30 (Na$_2$SO$_4$), 230 (NaF), 10 C$_{10}$H$_6$(HSO$_4$)$_2$, 0–0.8 pH=5.6–4.7	0.8–1.2	45–55	21	—	Good deposits	[36]
(NiSO$_4$), 200 (H$_3$BO$_3$), 30 (NaCl), 20 pH=5.8	0.15	30–32	16	—	Variation in sound pressure has no significant effect on structure of deposit	[16]

TABLE 19. Technological Conditions for Electrodeposition of Lead and Tin

Composition of the electrolyte, kg/m^3	Conditions of electrolysis		Ultrasonic conditions		Quality of the deposit	Source of data
	Current density, A/m$^2 \cdot 10^{-3}$	Temp., °C	Frequency, kc	Intensity, W/m$^2 \cdot 10^{-4}$		
[($C_6H_4OHSO_3$)$_2$Pb], 250 ($C_6H_4OHSO_3$), 25 No. 5 glue	0.5	20	21	5.8	Improved	[11]
($SnSO_4$), 50 (H_2SO_4), 100 (C_6H_5OH), 20	0.02-0.04	25	21	5.8	Improved	[11]

Chromium

The many investigations on the application of ultrasonic fields in the deposition of chromium indicate that ultrasonic vibration does not significantly improve the quality of the electrolytic deposits. Also, ultrasonic vibration inducing violent mixing of the electrolyte makes the electrolyte turbid and considerably increases the amount of suspended matter. A. M. Ozerov showed that electrodeposition of chromium in an ultrasonic field is accompanied by a small decrease (at low current densities) in the yield of the metal. This author assumed that ultrasonic vibration has a considerable effect on the electrochemical reduction of chromium ($Cr^{6+} \rightarrow Cr^0$) and facilitates and accelerates the reductions: $Cr^{6+} \rightarrow Cr^{3+}$ and $2H^+ \rightarrow H_2$.

The essential information on electrodeposition of chromium in an ultrasonic field is given in Table 15.

Cadmium

Electrolytic deposits of cadmium from solutions of sodium cyanide in an ultrasonic field always consist of small crystals. Ultrasonic vibration of high frequency and high intensity does not improve the brightness of the deposits. The conditions of deposition in an ultrasonic field are given in Table 16.

Iron and Cobalt

Electrodeposition of iron and cobalt from different electrolytes in an ultrasonic field at a frequency of 1200 kc produces deposits consisting of small crystals which are much brighter than those produced under ordinary conditions, and the corrosion resistance of the deposits increases by a factor of 3-4 as compared to that of deposits produced under ordinary conditions. The conditions of electrolysis are given in Table 17.

Nickel

The electrodeposition of nickel in an ultrasonic field has been investigated in great detail. The principal data on the electrochemical and ultrasonic conditions of deposition are given in Table 18.

Ultrasonic vibration has a strong effect on the electrodeposition of nickel and the effect depends to a considerable extent on the conditions of electrolysis. Thus, the effect of ultrasonic vibration on cathode polarization depends on the concentration of the principal salt in the electrolyte. R. Weiner found that the shift of the cathode potential toward positive values is insignificant (about 50 mV). A. M. Ginberg et al. also found that in electrolytes of type No. 1 the polarization is very little affected [67]. If the current density is increased from 2 to 10 A/dm^2, the polarization of the cathode increases by 250 mV. When nickel is deposited from

TABLE 20. Technological Conditions for Electrodeposition of Copper

Composition of the electrolyte kg/m³	Conditions of electrolysis		Ultrasonic conditions		Quality of the deposit	Source of data
	Current density, A/m²·10⁻³	Temp., °C	Frequency, kc	Intensity W/m²·10⁻⁴		
$(C_2H_5OH$, 10 cm³) $(CuSO_4)$, 200 (H_2SO_4), 30	0.25	–	330	–	Grain size increased	[62]
Acid electrolyte containing 0.3% copper	–	–	16	2	Improved	[38]
(H_2SO_4), 50 $(CuSO_4)$, 200	0.5	20	21	5.8	Good deposits	[11]
$(CuSO_4)$, 250	12.0	–	21	–	Improved	[36]
$(CuSO_4)$, 35–70 $(Na_4P_2O_7)$, 140–200 (NaH_2PO_4), 95–95 pH = 6.6	0.6–0.8	45	21.3	–	Improved	[36]
$(CuSO_4)$, 42.5 (H_2SO_4), 2	0.8	20	200–1000	≥1	Same quality as with mixing	[34]
Cyanide electrolyte	1.0	–	20	–	Improved	[59]
$(CuCN)$, 120 $(NaCN)$, 12–15 (Na_2CO_3), 10–25	2.0	40	16–20	–		[30]
$(CuCN)$, 60–80 $(NaCN)$, 12–20 $[(NH_4)_3PO_4]$, 5 $(CuSO_4)$, 200 (H_2SO_4), 60	1.6	30–60	15–25	0.5–1	Good deposits, anodes de-passivated	[37]
$(CuSO_4)$, 250 (H_2SO_4), 75	1.5–2.5	25–30	16	–	Structure improved, very high current densities	[26]

TABLE 21. Technological Conditions for Electrodeposition of Silver and Gold

Composition of the electrolyte kg/m³	Conditions of electrolysis		Ultrasonic conditions		Quality of the deposit	Source of data
	Current density, A/m²·10⁻³	Temp., °C	Frequency, kc	Intensity, W/m²·10⁻⁴		
(AgNO₃), 85-129	1.0-13.0	22	5200	4	Grain size decreased	[44]
K[Ag(CN)]₂, 62.5 (KCN), 20 (K₂CO₃), 15	0.2-1.0	20	34	0.3	Polarization decreased	[21]
Cyanide electrolyte			20		Good deposits	[59]
K[Ag(CN)]₂, 40 (KCN), 30.3 (K₂CO₃), 89.3	1.0-1.5	40	16	-	-	[30]
Electrolyte for gold plating contains 5·10⁻⁴ kg/m³ Au	0.2	20-50	16	2	Good deposits	[38]

electrolyte No. 2 the effect of ultrasonic vibration is quite different. The potential of the cathode improves, changing very little with increasing current density. A. M. Ginberg assumed that ultrasonic vibration renders the concentration gradient more uniform and has a specific effect on the formation of nickel hydroxide.

The kinetics of electrodeposition of metals of the iron group is often affected by the formation of metal hydroxides in the cathode layer. In an ultrasonic field, the current density at which hydrates begin to form in the cathode layer is shifted toward higher current densities. In this case the depolarization effect is particularly noticeable because it is due not only to a decrease in polarization but also to the absence of hydrates. However, hydroxides are formed even in an ultrasonic field in some electrolytes at high current density, but the grain size and porosity decrease. All these factors may slow down the discharge of nickel ions. It was found that the presence of hydroxide in an electrolyte subjected to ultrasonic vibration promotes an even larger shift of the cathode potential toward negative values than the addition of hydrosol in electrolytes not subjected to ultrasonic vibration.

Many investigators have noted that in a number of cases the yield of nickel is increased and the current density can be increased by a factor of 3-4 when ultrasonic vibration is used. R. Weiner obtained mirror deposits in an ultrasonic field using current densities of 3-8 A/dm², while under ordinary conditions the limiting current density did not exceed 1 A/dm².

Usually, ultrasonic vibration does not improve the mixing of electrolytes used for the deposition of nickel, and may even decrease it. The distribution of the metal on the cathode depends on the nature of polarization. Some investigators note that the bath functions better in its depth. The American investigator, T. Kozan, noted that the mixing of the Watt electrolyte used for the deposition of nickel is diminished by ultrasonic vibration. If the deposition is carried out under ordinary conditions, the throwing power decreases from 4.8 to 3% when the current density is increased from 5.5 to 10.8 A/dm². In an ultrasonic field the throwing power is decreased to 4.9%. Further increase

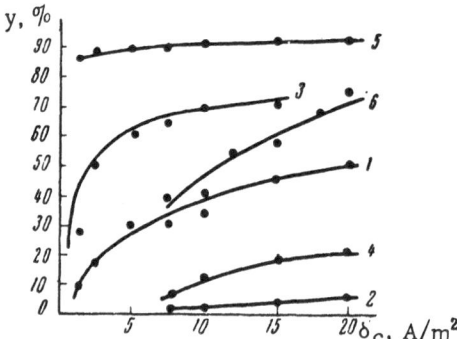

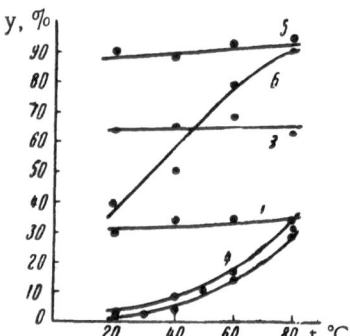

Fig. 33. Dependence of the composition of Zn—Co alloy on the current
density. Curves 1, 3, and 5 were obtained without ultrasound; curves 2, 4,
and 6 were obtained in an ultrasonic field.

Fig. 34. Dependence of the composition of the alloy on temperature.
t) Temperature, °C; y) amount of
cobalt in the alloy, %; 1-6 same
as in Fig. 33.

in the current density does not change the degree of mixing. B. Brown noted that the mixing effect of ultrasonic vibration makes the temperature of the electrolytic bath less critical. Good nickel deposits are obtained
at current densities 2-5 times higher in ultrasonic field than under ordinary conditions. When ultrasonic vibration is used in bright nickel plating one should not add high-molecular-weight compounds to the electrolyte,
since ultrasonic vibration may destroy them. Nickel plating of stainless steel in an ultrasonic field was carried
out in semi-industrial equipment. The rate of coating of the steel with nickel increased considerably, but preliminary cleaning of the surface to be plated could not be omitted.

It was found that the quality of the deposits is improved, the microhardness is increased, the deposits become brighter, and the internal stresses are decreased when deposits are made in an ultrasonic field. Also, nickel
deposits are less porous and more ductile. The acidity of the electrolyte during electrolysis is stabilized and the
composition of the electrolyte can be less complex. Ultrasonic vibration stabilizes the functioning of the anode.

Tin and Lead

The quality of tin and lead deposits is improved in an ultrasonic field. The deposits consist of small crystals and they are dense and less porous. The polarization of tin decreases. The function of the anode improves:
the anodes dissolve more uniformly and anodic slime is easily removed.

A study of the effect of ultrasonic vibration on the electrolytic deposition of tin and lead was made with
electrolytes containing sulfuric acid and phenolsulfonic acid. It was shown that high current densities can be
used. The effect of surface-active substances added to the electrolytes decreases.

The working conditions are given in Table 19.

Copper

The conditions for electrodeposition and the characteristics of ultrasonic vibration used during electrodeposition of copper are given in Table 20. Analysis of the experimental data shows that the acceptable current density is 6-8 times higher than those used without the ultrasonic field. The porosity of the deposits produced in an ultrasonic field is lower. Depending on the characteristics of the ultrasonic field, the copper deposits consist either of small or large crystals. Ultrasonic vibration has a particularly favorable effect on the
dissolution of anodes.

TABLE 22. Effect of Ultrasonic Vibration on the Values of the Constants A and B

Conditions of electrolysis		Without ultrasonic field		With ultrasonic field	
δ_C, A/m$^2 \cdot 10^{-2}$	t, °C	A	B	A	B
7.5	20	−0.28	1.28	−1.8	1.7
7.5	40	−0.24	1.28	−1.5	1.6
15	20	−0.8	1.26	−1.3	1.65

TABLE 23. Effect of Ultrasonic Vibration on the Constants A' and B'

Composition of the electrolyte, kg/m^3		Without ultrasonic field		With ultrasonic field	
Zn	Co	A'	B'	A'	B'
5	5	−1.02	0.8	−2.92	1.32
3	7	−0.2	0.53	−2.3	1.4
1	9	+ 0.74	0.26	−1.7	1.6

TABLE 24. Effect of Ultrasonic Vibration on the Constants A" and B"

Composition of the electrolyte, kg/m^3		Without ultrasonic field		With ultrasonic field	
Zn	Co	A"	B"	A"	B"
5	5	−0.07	80	6.25	2340
3	7	0.51	80	6.30	2300
1	9	1.23	80	6.49	1980

TABLE 25

Alloy	Without ultrasonic field		With ultrasonic field	
	A	B	A	B
Tl−Ni	−0.38	−0.56	1.07	−0.74
Tl−Co	−0.43	−0.60	2.0	−1.28

TABLE 26

Alloy	Without ultrasonic field		With ultrasonic field	
	A'	B'	A'	B'
Ni−Tl	−1.69	−250	−2.24	−800
Co−Tl	−1.75	−250	−0.22	−300

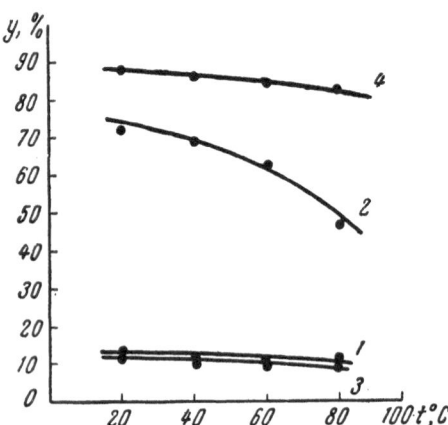

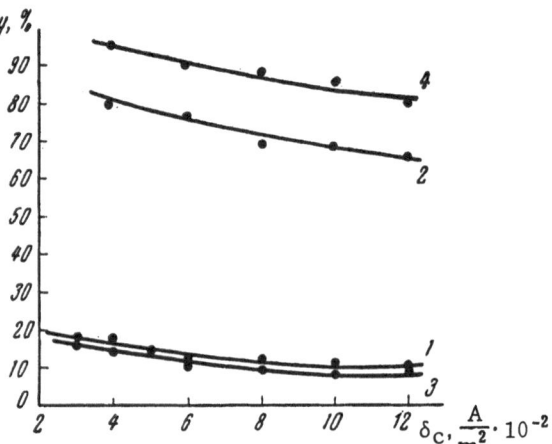

Fig. 35. Dependence of the compositions of Ni−Tl and Co−Tl alloys on the temperature of the electrolyte. 1, 2) Ni−Tl; 3, 4) Co−Tl. Curves 2 and 4 were obtained in an ultrasonic field.

Fig. 36. Dependence of the compositions of Ni−Tl and Co−Tl alloys on the current density (y is the amount of Tl in the alloy, %). 1-4) Same as in Fig. 35.

Silver and Gold

Electrolytic deposits of silver and gold produced in an ultrasonic field are of good quality. They consist of smaller crystals than those produced under ordinary conditions. The polarization decreases in an ultrasonic field. In the case of silver plating the current density can be increased to 1200 A/m². The yield in this case 85-96%. The deposition rate is 6-7 microns/min. The shape of the crystals changes in silver deposits produced in an ultrasonic field.

Some investigators noted that the adhesion of silver to the base metal is lowered by ultrasonic vibration when ordinary current densities are used. American and German scientists have used ultrasonic vibrations for gold plating of switches and circuit contacts. As the result of ultrasonic vibration the gold plate was almost free of impurities with low electrical conductivity. The conductivity of the gold-plated contacts was enhanced as a result.

The conditions of electrodeposition of silver and gold are given in Table 21.

Electrodeposition of Metal Alloys in an Ultrasonic Field

Ultrasonic vibration has a strong depolarizing action and, as a result, the hydrogen liberation potential and the metal separation potential are decreased, the limiting diffusion currents are increased, the yield is increased, and it is possible to carry out the electrolysis at higher current densities.

Consequently, ultrasonic vibration has a powerful effect on the processes of electrodeposition of alloys. It is well known that the composition of electrolytic deposits produced by simultaneous deposition of different metals depends to a great extent on such factors as the cathodic polarization of each of the metals, the diffusion current of the most inert component, and the hydrogen overvoltage on the alloy being deposited. Ultrasonic vibration affects the magnitude of these factors to a different degree for each metal and, consequently, affects the composition of the alloy deposited. The effect of ultrasonic vibration is also manifest in the change of the dependence of the composition of the deposit on the relative concentrations of the components of the electrolyte and on the conditions of electrolysis (current density, temperature, etc.).

There are relatively few publications dealing with the effect of ultrasonic vibration on the electrodepostion of alloys.

It was found that acoustic vibration affects the composition of brass deposits. For example, when the current density is $2 \cdot 10^2$ A/m^2, brass deposits produced without ultrasonic vibration contain 81% Cu, while the deposits produced in an ultrasonic field contain only 68% Cu.

In our laboratory G. R. Pobedimskii studied the effect of ultrasonic vibration on the electrodeposition of the following alloys: zinc–cobalt, nickel–thallium, and cobalt–thallium.

Zinc–Cobalt Alloys

It is well known that the equilibrium potentials of zinc and cobalt are close, and therefore they can be deposited simultaneously from solutions of their simple salts. Radioactive Zn^{65} was used in the investigation. The investigation showed that there is a simple relationship between the conditions of electrodeposition and the composition of the alloy deposited.

In this investigation the electrolytes were prepared by mixing two solutions in different proportions. The first solutions contained 47.88 kg/m^2 CoSO$_4 \cdot$ 7H$_2$O (10 kg/m^3 Co) and the second contained 44 kg/m^2 ZnSO$_4 \cdot$ $\cdot$ 7H$_2$O (10 kg/m^3 Zn). The pH was 2.5-2.0. The electrolysis was carried out in a glass container with a capacity of 100 cc. The container was placed in a glass jacket. The bottom of the glass container was glued to the bottom of the glass jacket, which was placed on an ultrasonic vibrator. Water from a thermostat was circulated continuously between the walls of the jacket and the glass containing the electrolytic bath so as to maintain the electrolyte at a constant temperature. A magnetostrictive vibrator was placed in the bath, where the temperature was kept at 20°C. The alloy was deposited on copper cathodes 1 × 2 cm. The cathode was placed vertically at the center of the bath. Platinum plates of the same size were used as anodes.

The magnetostrictive vibrator was driven by a UZM-1.5 generator with a nominal power rating of 1.5 kW. The frequency was 24 kc and the intensity was $0.3 \cdot 10^4$ W/m^2.

The composition of the electrolytic deposits was determined by the radioactive tracer method. The radioactive isotope Co^{60} (0.1 milliCurie/liter) was added to the cobalt solution.

The dependence of the composition of the electrolytic deposit on the value of the cathode current density and on the temperature of the electrolyte was studied for different ratios between the concentrations of the components in the solution. The experiments with ultrasonic vibration were alternated with experiments without ultrasonic vibration. The results of the investigation are shown in Figs. 33 and 34. (The Zn/Co ratio was 5 : 5 in electrolytes No. 1 and 2, 3 : 7 in electrolytes 3 and 4, 1 : 9 in electrolytes 5 and 6; the temperature was 20°C.)

The investigation showed that the dependence of the composition of the electrodeposited alloy on the concentration ratio in the electrolytes obeys the following equation:

$$\log \frac{P_1}{P_2} = A + B \log \frac{C_1}{C_2},$$
(1)

where P_1 and P_2 are the concentrations of cobalt and zinc in the electrolytically deposited alloy, in percent by weight, C_1 and C_2 are the concentrations of cobalt and zinc in the electrolyte, in g/liter, A and B are constants independent of the temperature and the current density. This equation was derived theoretically by E. M. Akhumov and B. Ya. Rozen. The values of the constants for the zinc–cobalt alloys in Eq. (1) were obtained experimentally and are given in Table 22.

The dependence of the composition of the electrolytically deposited zinc–cobalt alloy on the current density is represented by the following relationship:

$$\log \frac{P_1}{P_2} = A' + B' \log \delta_c,$$
(2)

where δ_c is the cathode current density, in A/m^2, A' and B' are constants independent of the current density.

The experimentally obtained values of the constants in Eq. (2) are given in Table 23.

The investigation showed that when the temperature of the electrolyte is 20°C the effect of ultrasonic vibration on the formation of the alloys consists of a sharp decrease in the concentration of cobalt in the electrolytic deposit. Apparently, ultrasonic vibration decreases the cathodic separation polarization of zinc to a greater extent than that of cobalt and, as a result, the concentration of zinc in the deposit increases.

It was found that the dependence of the composition of the electrolytically deposited alloy on the temperature of the electrolyte may be expressed by the following equation:

$$\log \frac{P_1}{P_2} = A'' - \frac{B''}{T},$$ (3)

where T is the absolute temperature of the electrolyte, A" and B" are quantities independent of the temperature.

The values of the constants in Eq. (3) are given in Table 24.

Equation (3) can be obtained by transforming the equation representing the dependence of the current density on the temperature proposed by S. V. Gorbachev. The coefficient B" is a linear function of the difference between the effective activation energies of electrochemical deposition of the components of the alloy. The abrupt increase in the coefficient B" as the result of ultrasonic vibration is due to the difference in the decrease of the separation activation energies of zinc and cobalt.

Dense, matte, gray deposits 5-20 microns thick were obtained in this investigation. The yield did not exceed 40%.

Nickel—Thallium and Cobalt—Thallium Alloys

The equilibrium potentials of thallium, nickel, and cobalt are sufficiently close so that they can be deposited simultaneously from simple solutions of their salts.

The nickel—thallium alloy was deposited from an electrolyte with the following composition: 238.4 kg/m^3 NiSO$_4 \cdot$ 7H$_2$O, 3.08 kg/m^3 Tl$_2$SO$_4$, 20 kg/m^3 H$_3$BO$_3$, 100 kg/m^3 Na$_2$SO$_4 \cdot$ 10H$_2$O. The pH was 2.5. The investigation was made by the radioactive tracer method; 0.1 mC/liter of the Tl^{204} isotope was added in the form of thallium sulfate.

The thallium—cobalt alloy was deposited from a solution with the following composition: 238.4 kg/m^3 CoSO$_4 \cdot$ 7H$_2$O, 3.08 kg/m^3 Tl$_2$SO$_4$, 20 kg/m^3 H$_3$BO$_3$, 100 kg/m^3 Na$_2$SO$_4 \cdot$ 10H$_2$O. The pH was 2.5.

Radioactive Tl^{204} (0.1 mC/liter) and radioactive Co^{60} (0.9 mC/liter) were added in the form of sulfates.

Since Tl^{204} disintegrates with the emission of high-energy β particles (0.785 meV) and without the emission of γ radiation, while the Co^{60} isotope emits γ radiation, the concentration of thallium in the electrolytically deposited alloy was determined by measuring the β activity of the deposit, while the concentration of cobalt was determined by the γ activity of the deposit. The fact that one can determine simultaneously the concentrations of both components of the alloy (thallium and cobalt) makes it possible to use this method to determine the composition of the electrodeposited nickel—cobalt—thallium alloy.

The dependence of the composition of electrolytically deposited alloys on the magnitude of the cathode current density and on the temperature of the electrolytes were also investigated. The experiments in an ultrasonic field were alternated with experiments without the ultrasonic field. The results of the experiments are shown in Figs. 35 and 36. The investigation showed that ultrasonic vibration sharply increases the concentration of thallium in both alloys and that this increase is more pronounced in the cobalt—thallium alloy. The compositions of the alloys produced under ordinary conditions differ very little, while for the alloys produced in an ultrasonic field the cobalt—thallium alloy contains much more thallium than the nickel—thallium alloy. The experimentally determined constants A and B in the equations given above for the current density are given in Table 25 for the case of electrodeposition of Ni—Tl and Co—Tl alloys.

Ultrasonic vibration sharply increases the value of the constant A and somewhat decreases that of the constant B, which characterizes the slope of the curve.

The values of the constants A' and B' in the equation for the temperature obtained in this investigation are given in Table 26.

Ultrasonic vibration decreases the angular coefficient B' in the equation, and this decrease is more pronounced in the case of the Ni−Tl alloy than in the case of the Co−Tl alloy.

The change in the composition of the alloys and also the change in the character of the dependence of the composition on the temperature resulting from the effect of ultrasonic vibration on electrodeposition are apparently due to a decrease in polarization and effective activation energies for electrochemical deposition of the components of the alloy.

Ni−Tl plating containing no more than 20% Tl, produced by electrodeposition, is strong, uniform, and bright; the thickness is as much as 10 microns. Electrolytic deposits of nickel containing the radioactive thallium isotope Tl^{204} can be used as a source of β radiation. One can prepare standards of β activity by depositing an alloy of nickel and radioactive Tl^{204}.

The cobalt−thallium deposits produced under ordinary conditions are not of good quality. Ultrasonic vibration affects the electrodeposition process by increasing the concentration of Tl in the deposit, but this does not improve the quality of the deposit. The maximum increase in the amount of thallium in the deposit occurs when the ultrasonic intensity is low. An increase in ultrasonic intensity from $0.2 \cdot 10^4$ to $0.5 \cdot 10^4$ W/m^2 decreases the amount of thallium in the deposit.

CHAPTER III

EFFECT OF ULTRASONIC VIBRATION ON AN ANODIC SOLUTION

The effect of ultrasonic vibration on an anodic solution of a metal has been studied much less than the effect of ultrasonic vibration on the electrodeposition of metals. However, knowledge of the effect of ultrasonic vibration on anodic solution of metals is very important for industrial purposes. There are data indicating that ultrasonic vibration increases the rate of anodic solution and leads to breaking up of the anode when the intensity of vibration is very high ($10 \cdot 10^4$ W/m^2). The effect of ultrasonic vibration on the anode overvoltage was studied in [49], where it was shown that anode overvoltage decreases under the effect of ultrasonic vibration in the case of a solution of iron, copper, and cadmium. The results obtained in this investigation are shown in Fig. 37.

A similar investigation was made in [50], where the effect of an ultrasonic field on the anodic solution of copper was studied. The author of this investigation used a quartz vibrator providing vibrations with a frequency of 710 kc and an intensity of $2 \cdot 10^4$ W/m^2, and found that when the current density is of the order of $2 \cdot 10^2$ A/m^2 or higher the ultrasonic field decreases the polarization of the anode. Ultrasonic vibration has no effect at lower current densities. The anode becomes bright after prolonged polarization in an ultrasonic field. The products of anodic solution accumulate at the crests of the ultrasonic waves, while at the troughs the anodes become bright.

At the laboratory of the Kazan Chemicotechnological Institute we investigated anodes of different structures: a polycrystalline copper sample of 99% pure cast copper, a copper single crystal in which different faces emerged at the surface of the crystal, and also anodes of cast, rolled, and electrolytic nickel. All the experiments were made in an ultrasonic field with a frequency of 21 kc and an intensity of $3 \cdot 10^4$ W/m^2. The electrolyte contained 200 kg/m^3 CuSO$_4 \cdot$ $\cdot$5H$_2$O and 60 kg/m^3 H$_2$SO$_4$ [51].

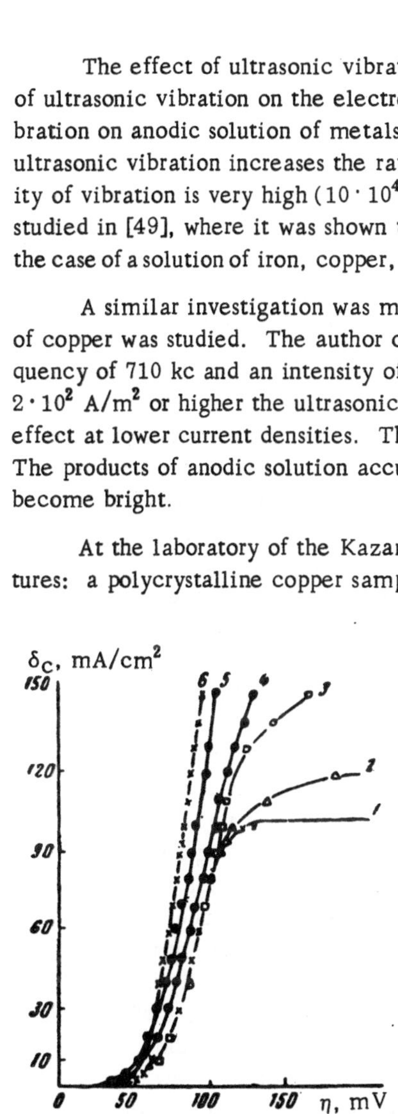

We investigated the rate of solution of the anode, the uniformity of solution of the anode, and the change in the surface hardness of the anode in an ultrasonic field.

Rate of Solution of the Anode

Figure 38 shows the experimental results concerning the variation in solution rate of the anode of polycrystalline copper in an ultrasonic field. The measurements were made at different current densities (from 200 to 500 A/m^2).

The investigation showed that the rate of solution of nickel anodes in an ultrasonic field is higher than that of copper anodes. Apparently, this is due to the fact that ultrasonic vibration has a greater effect on the solution rate of small and less ordered crystals. In fact, the solution rate of single crystal copper anodes in an ultrasonic field is 10% different than under ordinary conditions, while the rate of solution of polycrystalline anodes increases 25%. The study of the electrolytic solution of nickel showed that under ordinary conditions cast polycrystalline anodes dissolve more rapidly than electrolytic or rolled anodes. The solution rate is accelerated by ultrasonic vibration.

Fig. 37. Variation of anodic overvoltage during deposition of copper in an ultrasonic field. The experimental conditions were: $f = 960$ kc; I = $3 \cdot 10^4$ W/m^2; the electrolyte consisted of a 3 N H$_3$PO$_4$ solution and a 2 N Cu(PO$_4$)$_2$ solution. 1) Without ultrasonic field; 2) mixing at 1000 rpm; 3) mixing at 2000 rpm; 4) mixing at 3000 rpm; 5) mixing at 6000 rpm; 6) in an ultrasonic field.

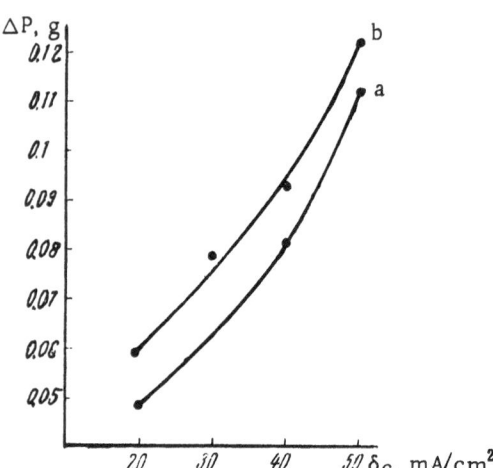

Fig. 38. Effect of ultrasonic vibration on the rate of dissolution of anodes (ΔP is the change of weight of the anode during dissolution). a) Under ordinary conditions; b) in an ultrasonic field.

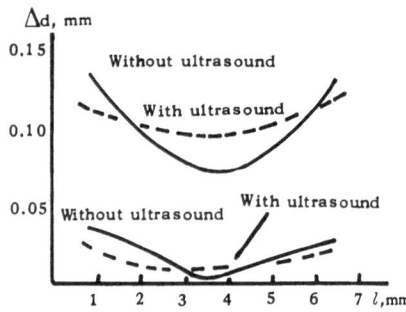

Fig. 39. Effect of ultrasonic vibration on the uniformity of dissolution of the anode (ΔP is the change in thickness of the electrode during electrolysis; l is the width of the anode). a) Under ordinary conditions; b) in ultrasonic field. 1) Time of electrolysis, 4h; 2) time of electrolysis, 0.5 h.

Cast anodes dissolve four times more rapidly in an ultrasonic field than under ordinary conditions. The solution rate of rolled and electrolytic anodes increases by a factor of two or three in an ultrasonic field. Under ordinary conditions the rate of solution of rolled anodes is the slowest, and ultrasonic vibration has little effect on the rate of anodic solution. In this case the crystallographic structure and the degree of order of the structure of the anodes play very important roles. Metallographic and x-ray analyses showed that cast nickel anodes are more texturized. All this leads to slower solution rates for rolled anodes under ordinary conditions and to the fact that ultrasonic vibration has less effect on the solution of rolled than on the solution of cast anodes. The crystallographic structures of rolled anodes resemble that of single crystals because rolled anodes are highly texturized. Copper anodes made of single crystals dissolve much more slowly than polycrystalline anodes. The effect of ultrasonic vibration on single crystal anodes is affected very little by ultrasonic vibration.

The fact that the solution rate of the anodes is accelerated by ultrasonic vibration indicates that the anodes are less passivated. This is particularly characteristic of nickel. Thus, under ordinary conditions the anodes become covered with a dark passive film after a few minutes of electrolysis and the intensity of the current in the circuit drops. In an ultrasonic field this film is either completely absent or appears after a longer period of time at higher current densities. The increased solution rate of the anodes in an ultrasonic field and the resulting depassivation can be explained by the strong mixing and dispersing effect of ultrasonic vibration. The mixing of the electrolyte renders the concentration gradients more uniform and promotes the removal of the products of anodic solution from the electrode layer; foreign particles are removed from the metal surface as the result of the dispersing effect. The dispersing effects prevent the formation of oxide films on the anode, i.e., they activate the surface of the anode. Depending on the conditions of electrolysis and the nature of the metal, the process of solution will be affected essentially either by mixing or by dispersion. For example, in the case of a solution of copper where the change in concentration of the electrolyte is small, i.e., the concentration polarization is low, mixing promotes the removal from the anode of different particles

formed as the result of the dispersion and solution of the anode. As a matter of fact, the yield measured with a copper coulometer is higher in an ultrasonic field than under ordinary conditions (over 100%), and this indicates that the anode is dispersed. We ran control experiments to determine the dispersing effect of ultrasonic vibrations. A copper sample was placed at the center of an ultrasonic beam and was subjected to vibration for 30 min without application of current. During this time the weight of the sample increased by 0.006 g. Therefore, the apparent yield of the anode becomes equal to 116%.

Unlike the dissolution of copper, the solution of nickel is accompanied by other phenomena. The anodic yield during electrolytic solution of nickel in an ultrasonic field increases, but this is not due to the dispersing effect alone. In fact, the experiments showed that after 30 min of ultrasonic vibration without application of

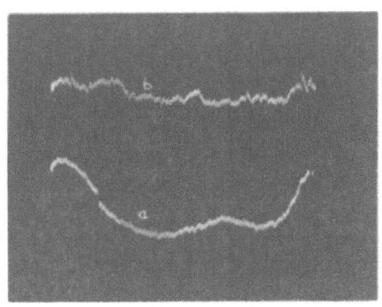

Fig. 40. Autoradiograph of the anode surface. a) Distribution of the metal on the surface of the anode after dissolution under ordinary conditions; b) distribution of the metal on the surface of the anode after dissolution in an ultrasonic field.

current, only 0.004-0.008 g of nickel is dissolved. The rate of solution of cast anodes is highest. The difference between the amounts of metal dissolved under ordinary conditions and in an ultrasonic field after 30 min of electrolysis is almost 0.1 g, i.e., much greater than that resulting from the dispersing effect. All this indicates that the acceleration of the solution of nickel anodes in an ultrasonic field is due essentially to a decrease in anodic polarization and to depassivation of the surface.

Thus, the study of the anodic solution of nickel and copper showed that ultrasonic vibrations can increase the solution rate of anodes as the result of the dispersing effect and also as the result of the decrease in concentration polarization. In the case of solution of copper anodes, where the change in concentration is small, the dispersing effect is the dominant factor. Electrolytic solution of nickel is accelerated essentially as the result of electrochemical changes, i.e., decrease of anodic polarization and depassivation of the surface.

Uniformity of Solution of Metals on the Surface of the Anode

Figure 39 shows the degree of uniformity of the solution of anodes in an ultrasonic field as compared to that under ordinary conditions. Measurements were made with an optimeter at three points along the median of the sample. It was shown that in an ultrasonic field the anode is dissolved uniformly over the whole surface (and the longer the time of electrolysis, the more uniform the solution).

We used the autoradiographic method to determine more precisely the uniformity of the solution of the anode. The experiments confirmed even more clearly the results obtained with the optimeter. The photomicrograph shown in Fig. 40 illustrates the experimental results (the width of the sample is represented along the x axis; the degree of uniformity of the distribution of the metal is plotted on the y axis). The layer of copper deposited on a radioactive base had a thickness of 50 microns. The uniformity of the distribution of copper over the surface was checked and the sample was then used as the anode. The experimental conditions were as follows: $\delta = 200$ A/m^2; t = 30°C; t = 15 min; $f = 23$ kHz; I = 3·10^4 W/m^2.

Effect of Ultrasonic Vibration on the Mechanical Properties of Anodes

The accelerating effect of ultrasonic vibration on the solution of anodes may affect such mechanical properties of the anodes as hardness and internal stress. Probably ultrasonic vibration partially relieves the stress-strain state in the surface layer of the metal. As the result, the constants of the crystal lattice change. X-ray diagrams of polycrystalline copper samples, before and after the experiments, were made with an RKÉ camera. The interference fringes corresponding to the sample treated in an ultrasonic field are shifted somewhat toward larger values of the lattice constant, i.e., toward the equilibrium state of the crystal lattice.

The constant of the crystal lattice of the anode subjected to ultrasonic vibration for 4 h changes on the average by 0.2%. This change is due to a decrease in the elastic stresses due to rolling of the copper anodes. The value of elastic stresses can be calculated by the Sekito formula:

$$\tau = \frac{\Delta dE}{d\sigma},$$

where Δd is the change of the lattice constant, d is the lattice constant, E is Young's modulus, and σ is the Poisson coefficient.

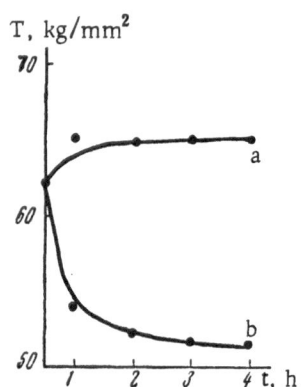

Fig. 41. Variation of the micro-hardness of the surface of the anode during electrolysis. a) Under ordinary conditions: b) in ultrasonic field.

In this case ultrasonic vibration acts in a manner similar to low-temperature heat treatment and restitution occurs as a result. Ultrasonic vibration supplies additional energy to the atoms of the deformed crystal. As a result, the atoms are regrouped and the structure of the crystal approaches the structure of the undeformed equilibrium state. Internal stresses decrease by 62 kg/mm². In our experiments restitution was not completed, since the crystal lattice did not reach the equilibrium state with a constant equal to 3.607 Å.

The fact that processes of restitution do occur in an ultrasonic field is confirmed by the changes in mechanical properties. The surface hardness of samples subjected to ultrasonic vibration changes to a considerable extent. Figure 41 shows that the surface hardness of copper anodes dissolving under ordinary conditions remains almost unchanged during electrolysis. However, after the anodes have been dissolved in an electrolyte subjected to an ultrasonic field their surface hardness decreases sharply during the first minutes of electrolysis and then remains almost constant. The change in surface hardness of copper anodes subjected to ultrasonic vibration is about 18% as compared with that of their initial hardness; the change is 30% for nickel anodes. These results lead us to assume that the structure of nickel anodes undergoes greater changes than that of copper anodes under the influence of ultrasonic vibration.

However, these structural changes occur only in the surface layer of the metal.

It was noted earlier that the crystal structure is a basic factor determining microhardness. However, the microhardness can also depend on other factors. In our case the decrease of internal stresses in the anode and the increased perfection of the crystal structure are the main factors responsible for the decrease in microhardness of anodes subjected to ultrasonic vibration, since the grain size remains unchanged.

The acceleration of the rate of dissolution of anodes is economically disadvantageous, since a greater amount of metal is spent as the result of breaking up of the anode. However, the use of ultrasonic vibrations is apparently advantageous when the anode is passivated during electrolysis. A properly chosen ultrasonic field may induce depassivation of the anode surfaces. When anodes are not passivated during electrolysis they should be screened from the intense ultrasonic beam so as to avoid breaking up of the anodes.

A. I. Ryazanov, A. I. Vol'fson, and E. O. Chigranova studied the effect of ultrasonic vibration on the anodic solution of palladium in a 6 N HCl solution and found that ultrasonic vibration with an intensity of $2 \cdot 10^4$ W/m² leads to considerable anodic depolarization during anodic solution of palladium. The ultrasonic vibration intensifies the anodic solution of palladium and it becomes possible to obtain a concentrated solution of palladium chloride (of the order of 500 kg/m³, instead of the 300 kg/m³ obtained by anodic solution without ultrasonic vibration). The authors assume that the depolarizing effect of ultrasonic vibrations on anodic solution of palladium in strong acids can be explained by the mixing and dispersing action of ultrasound. This action detaches the layer of salt formed on the surface of the anode and, as the result, further solution becomes possible, and the diffusion of the products of electrolysis from the anode into the solution is accelerated [52].

The rapid removal of the anodic film, the concentration equalization of the electrolyte about the anode, and the decrease in mechanical strength result in a more stable process, so that the metal is dissolved more uniformly over the surface of the anode. This result must be taken into account in determining the practicability of ultrasound in technological processes.

Breaking up of the nickel anode was also observed in [11]. The nickel powder accumulated at the bottom of the electrolytic bath. The study of the electrodeposition of chromium showed that the heavy film formed on the surface under ordinary conditions is rapidly removed from the surface into the electrolyte and, as a result, the electrolyte becomes turbid, with particles suspended in it. The electrolyte also became turbid in the case of electrodeposition of chromium due to the formation of suspensions of insoluble chromium salts of lead.

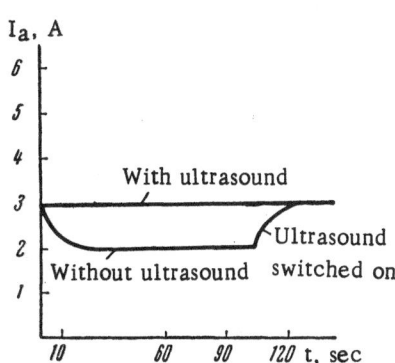

Fig. 42. Variation of the anodic current with time during electrodeposition of cadmium in an ultrasonic field.

The use of ultrasonic vibration in industrial nickel plating and copper plating has shown that the anodes work very well without the addition of depassivators to the electrolyte. The depassivating effect of ultrasonic vibrations has a strong effect on the function of the anodes. It was shown in [53] that the metals which are passivated in the solution begin to dissolve under the influence of an ultrasonic field. The variation of the anodic current with time under the influence of an ultrasonic field is shown in Fig. 42.

At the same time, a number of investigators have shown that ultrasonic fields do not always eliminate passivity. For example, in the case of an anodic solution of iron in dilute alkali the anode is not activated even when chlorides are added to the solution. Apparently intense mixing suppresses the activating effect of the chlorine ions under these conditions. Acceleration of the passivation of anodes in an ultrasonic field was also found in the case of nickel in a 0.4 N Na_2SO_4 solution [10].

In [10] the authors studied the effect of ultrasonic vibration on the solution of metals and came to the conclusion that the acceleration or deceleration of the solution of metals in an ultrasonic field is the result of intense mixing due to the ultrasonic vibrations. Ultrasonic vibration facilitates solution in cases where the accumulation of the solution products in the electrode layer prevents further dissolution of the metal. It is also possible that ultrasound promotes removal of the layer of gas from the surface of the dissolving metal, this layer normally preventing contact between the metals in the dissolving medium in the case of solution without the application of current.

If, however, ultrasonic vibration removes from the surface of the metal substances which promote the solution of the metal (chlorine ions, for example, in the case of solution of iron or nickel), then solution is slowed down by ultrasonic vibration. The facilitation of the evolution of oxygen on the surface of the anode as the result of ultrasonic field (in the case of solution of nickel, for example) may also facilitate solution.

Aside from all this, ultrasonic vibration can also break up the metals, and metal particles chemically bound to oxygen atoms are detached from the surface and are removed into the solution. This also promotes activation of the metal surface and eliminates its passivity [10]. Apparently, one of these factors is dominant, depending on the nature of the metal, the conditions of electrolysis, and the intensity of the ultrasonic field.

Anodic Oxidation of Aluminum in an Ultrasonic Field

The effect of ultrasonic vibration on anodic oxidation of aluminum has been studied by many investigators. It was shown in [19] that anodic films formed in an ultrasonic field with a frequency of 900 kc are not strong. The films are brittle and detach themselves from the base metal in the form of flakes. These experiments were made in an ultrasonic field inducing strong cavitation, which affects the electrode processes.

A. F. Bogoyavlenskii and V. A. Kochergina oxidized aluminum in a field with an intensity of $1.5 \cdot 10^4$ W/m^2 and a frequency of 23 kc. In these experiments the authors determined more precisely the role of oxygen in the formation of the surface film and the effect of changes in the concentration at the electrodes. The ultrasonic equipment used in this investigation was adapted for physicochemical investigation: the vibrator was of a magnetostrictive type with a surface of 100 cm^2. The electrolytic bath consisted of a glass tank with a capacity of 200 cc. The possibility of using lead, nickel, and steel cathodes for anodizing was investigated. Aluminum samples 2×4 cm were made from DTM-16 aluminum. Both sides of the samples were anodized. The anodic films were tested for stability, porosity (by the oil extraction method), brittleness (tested by measuring the microhardness), thickness of the film [determined gravimetrically and by direct measurements with a precision optical instrument (optimeter)]. The thickness of the film was determined after it was removed from the base metal. All the experiments were made in a 20% sulfuric acid solution (the current density was

TABLE 27. Effect of Ultrasonic Vibration on the Rate of Scale Removal from Steel

Steel No.	Steel descaling time, min			Increase in descaling rate, %
	With application of electric current; $\delta_c = 800$ A/m^2	Application of ultrasound without current	Simultaneous application of electric and ultrasonic fields, $\delta_c = 300$ A/m^2	
ÉYal-T	13	13.5	8	160
ÉYal-Nb	15	16	6	250
ÉI-100	19	17	8	240
ÉI-475	17	14.5	5	340
3KhV-8	20	16	4.5	440
Kh-12	18.5	18	4	460
65-7	18	19.5	8	220
ST-20	21	20	8	260
U-10	24	25	9	260

$1 \cdot 10^2$-$5 \cdot 10^2$ A/m^2, the temperature was 20-21°C, and the time of electrolysis was 20 min). First, lead cathodes were tried. However, they become rather rapidly covered with a black small-grain deposit which was shaken off the cathode by the ultrasonic vibration and accumulated at the bottom of the electrolytic bath. Thus, lead cathodes cannot be used. All the other experiments were made with steel cathodes. The change from lead and nickel cathodes to steel cathodes did not produce any significant change in the potential difference between the electrodes. The experimental results show that in all cases ultrasonic vibration decreases the porosity somewhat, and increases the hardness and brittleness of the anodic films. The films produced in an ultrasonic field had a high resistance to corrosion.

The kinetics of anodization at a current density of 10^2 A/m^2 is not changed by ultrasonic vibration to any significant degree. Good oxide films are obtained when the current densities are 3-5 · 10^2 A/m^2 without any complications induced by high current densities under ordinary conditions.

Whitish areas are formed at the troughs of the ultrasonic waves in the case of anodization of DTM-16 aluminum. These areas interrupt the uniformity of the structure of the films; they are of a different color. These areas are eliminated when the ultrasonic field intensity is decreased or if the sample to be anodized is placed outside the focus of the wave.

Cleaning, Degreasing, and Descaling in an Ultrasonic Field

The preparation of the surface plays a very important role in the electrodeposition of metals. At the present time, ultrasonic vibration is widely used for degreasing articles to be electroplated [16, 54, 55]. Ultrasonic vibration is very effective in these operations and makes it possible to speed up cleaning processes considerably, improves the quality of the surface, replaces undesirable operations (e.g., sandblasting), lowers the expense of washing materials, and improves working conditions.

Ultrasonic vibration accelerates the movement of the liquid near the diffusion layer, induces cavitation phenomena, and emulsifies oil films. Regardless of the type of cleaning, the principle of ultrasonic cleaning consists in intense motion of the liquid by ultrasonic vibration. When the intensity of the ultrasonic field is low (of the order of $5 \cdot 10^3$ W/m^2) degreasing and cleaning are slow and the vibration is not very effective. At ultrasonic field intensities ranging from 5 to $10 \cdot 10^4$ W/m^2 cleaning is rapid and provides good surfaces [54]. An increase of the intensity from 20 to $30 \cdot 10^4$ W/m^2 does not accelerate the cleaning processes because violent cavitation prevents proper distribution of the ultrasonic vibrations; thus, the accleration of the liquid flow is impaired and the sound pressure decreases [56]. Therefore, one must select adequate equipment for cleaning and degreasing. Also, one must take into account the singularities of the different cleaning operations due to the type and properties of impurity to be removed and the material to be cleaned. Consequently, optimum technological conditions must be determined experimentally, depending on the material and the size of the articles to be cleaned. Grease and solid particles are removed in aqueous solutions of acids, alkalis, and organic solvents. But it should be noted that cavitation is much more difficult to induce in organic solvents and that the erosion due to cavitation is much less strong. Therefore, nonfatty impurities should be removed in aqueous solutions.

The compositions of cleaning solutions and the cleaning conditions used in industry have been investigated in great detail [16, 53, 54, 55].

The conditions for the removal of scale from machine parts have also been described by many investigators. We present here only the data on the effect of ultrasound on the removal of scale from steel parts in different media [57].

The experiments were made with the ultrasonic equipment used for physicochemical investigations. The results of the experiments are given in Table 27. These results lead us to the following conclusions.

Ultrasonic vibration destroys scale on steels in a 10% HCl solution. Depending on the type of steel, the scale is completely removed after 13-25 min. Ultrasonic vibration increases the rate of electrochemical removal of the scale. The process of removal of the scale from samples in a 10% HCl solution is accelerated by a factor of 2-2.5 by ultrasonic vibration and by a factor of 3-4 or more in a 10% H_2SO_4 solution. It was found that the low-alloyed steels are attached much more in an ultrasonic field than highly alloyed steels, which are more resistant.

Application of Ultrasonics for Electroplating Processes in the USA

Earlier, ultrasonic vibration of high frequency was recommended in the USA to increase the rate of electrochemical processes, but at the present time equipment operating at a frequency of 20-40 kc is recommended. Ultrasonic vibration must satisfy the following requirements: the intensity of ultrasonic vibration should not exceed 10^4 W/m^2, since a higher intensity would induce cavitation, which may damage the electrolytically deposited plating. Experiments with pulsed ultrasonic fields have been made in the USA. These experiments showed that the danger of damaging the deposits by cavitation, even by damped ultrasonic vibration of high intensity, is eliminated completely, the quality of the deposits improved (hardness, strength, uniformity), and the rate of electrodeposition increased. Unsatisfactory results obtained at the beginning were attributed to the high frequencies and high intensities of the ultrasonic vibrations.

At the present time it is emphasized that the advantages of ultrasonic vibration are obtained only in low-frequency ultrasonic fields of 20-40 kc.

It should be noted that, according to American scientists, the accelerating effect of ultrasonic vibration on electrodeposition of metals is not due merely to improved mixing, but is much more complex.

American scientists report that the main advantage of ultrasonic vibration during nickel plating is that the control of the temperature of the bath need not be very precise. At the same time, the current density can be increased by a factor of 5. These scientists ascertained that in some cases the current density can be increased by a factor of 30. For example, under ordinary conditions nickel plating is carried out at current densities no higher than 27 A/m^2, while in an ultrasonic field with an intensity of $3 \cdot 10^3$ W/m^2 the current density can be increased to 400 A/m^2. In the case of cadmium plating an increase of the current density above 2-3 A/dm^2 under ordinary conditions renders the plating unsatisfactory, while in an ultrasonic field the current density can be increased to $8 \cdot 10^3$ A/m^2, and under these conditions the rate of deposition is 5 microns per min. The rate of chromium plating increases by a factor of 5 in an ultrasonic field and that of silver plating increases by a factor of 15. Also, these platings can be deposited on intricate surfaces.

The characteristic of all platings deposited in an ultrasonic field is better adhesion resulting from better cleaning of the surface. In the case of zinc plating in drums ultrasonic vibration increases the plating rate by a factor of 2. For silver plating deposited from cynaides the current density can be increased to 150 A/m^2 without impairing the quality of the deposits. The adhesion also improves. Thus, ultrasonic vibration apparently has a beneficial effect on the deposition of silver plating.

Although these and other results mentioned are not final, they do indicate that ultrasonic vibration is very promising in the electrodeposition of metals. In connection with this, one should mention the following communications from the USA. For example, it is indicated that ultrasonic vibration can be used in anodization and that the current density in the case of electrolytic baths containing sulfuric acid can be increased, hence the quality of the plating is improved.

At the present time, large ultrasonic equipment is being used in the USA for the preparation of surfaces for electrodeposition, as well as for the actual electrodeposition.

In modern American plants surfaces are prepared for electrodeposition with ultrasonic equipment operating continuously and using magnetostrictive transducers. Trichlorethylene is used as the solvent; only 5-8 liters of the solvent is used per ton of metal. The complete cycle takes 6 min, of which cleaning takes 30 sec.

Similar continuous ultrasonic equipment is used in the USA to prepare steel tape for chromium plating. This equipment is very economical.

Etching used to prepare surfaces for electroplating is also done in an ultrasonic field in the USA.

Including the preliminary mechanical cleaning, some of the high-power continuously operating ultrasonic equipments are as long as 36 meters; as much as 15 tons of tape is cleaned at one time.

In the USA it is believed that this type of cleaning equipment has a promising future, particularly for removing grinding compounds from metal surfaces to be plated, since these compounds are difficult to remove by other methods.

CHAPTER IV

BRIEF OUTLINE OF ULTRASONIC EQUIPMENT IN USE TODAY

Sonic and ultrasonic vibration in a wide range of frequencies is being employed on an increasing scale in various branches of industry. There are different methods of producing ultrasonic vibrations. A detailed description of the principles of operation of ultrasonic equipment can be found in several texts concerning the subject [4, 5, 8, 16]. In practice, only magnetostrictive and piezoelectric transducers have been used.

Several magnetostrictive vibrators of the A615.01, A615.05, A615.06, and A615.04 types are shown in Fig. 43. The plates of the vibrators are usually made of nickel, permalloy, and permendur, i.e., materials with a strong magnetostrictive effect and large breaking strength.

The vibrating surfaces of these vibrators differ. The A615.06 vibrator has a surface of 16 cm^2, the A615.04 vibrator has a surface of 70 cm^2, the A615.05 vibrator has a surface of 85 cm^2, the A615.01 vibrator has a surface of 90 cm^2, etc. The surface area can be increased to 200 cm^2 by using removable diaphragms.

With these vibrators the frequency can be varied from 16 to 200 kc, the intensity being $1 \cdot 10^5$ W/m^2.

The second widely used method of producing ultrasonic vibration is based on the piezoelectric effect. Quartz and barium titanate are mainly used in these apparatuses. The ultrasonic frequency produced with these apparatuses is of the order of several megacycles. Figure 44 shows a diagram of such a vibrator.

In both types of apparatus the vibration is excited by high-frequency electric generators. The ultrasonic generators used in laboratories and in industry are shown in Figs. 45 and 46.

The UZG-10 generator is constructed so as to drive several vibrators and is provided with four PMS66 vibrators. The efficiency of such a generator is at least 70%.

Different types of ultrasonic generators are used in scientific research. Brief data on the equipment used in laboratories and in industry are given in Table 28.

Construction of the Electrolytic Bath and Position of the Magnetostrictor

The position of the vibrator with respect to the cathode is very important in obtaining good-quality metal deposits in an ultrasonic field.

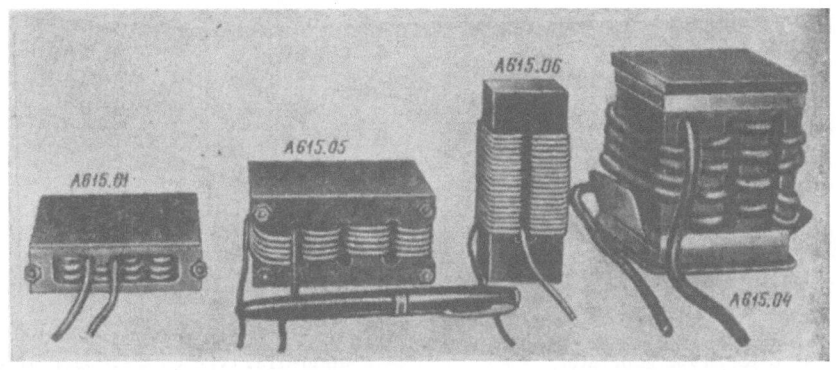

Fig. 43. View of magnetostrictive vibrators of the A615 type.

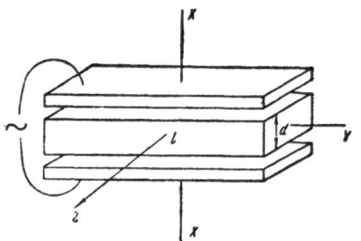

Fig. 44. Orientation of the piezoquartz plate between two vibrating electrodes (l is the length of the plate, d is the thickness of the plate).

In laboratories small electrolytic baths made of glass, plastic, or metals are used. Thus, the electrolytic bath can be placed directly on the surface of the magnetostrictor so that the ultrasonic vibrations propagate in the direction parallel to the cathode. To decrease acoustic losses, the walls of the bath must be thin and the bottom must be in close contact with the surface of the magnetostrictor. If one uses vibrations from one side only, one must place a reflecting surface opposite the vibrator; this reflector often consists of a layer of porous rubber 0.5-1.0 cm thick.

A number of investigators have used plexiglass or vinyl baths with the vibrators mounted in the bottom of the bath. Membranes are often soldered to these vibrators to increase the emitting surface. A bath of this type is shown in Fig. 47.

Different types of baths are used in industry. The most widely used are the 10-150-liter UZV baths made of one piece of metal and lined with vinyl. The baths used differ in size and in the type of PMS vibrator used. Thus, the UZV-15 bath has a capacity of 35 liters and uses the PMS-6M vibrator. The UZV-18 bath has a capacity of 150 liters and has four vibrators.

Baths in which NÉL-IV vibrators were mounted in different places, depending on the capacity of the bath, were constructed at the Scientific Research Institute of the Automobile Industry. Usually, the vibrators are placed so that the maximum amount of acoustic energy arrives at the cathode, while the anode receives only scattered energy. In a bath with a capacity of 7-10 liters the vibrator is placed under the articles to be plated and the field is directed from bottom to top. In baths with a capacity greater than 70 liters the field is perpendicular to the support rod at the end wall below the articles to be plated. In baths up to 1200 liters the field is parallel to the support rod and is directed toward the articles to be plated.

Fig. 45. Laboratory type generator.

Fig. 46. General view of the industrial type UZG-10 generator.

Fig. 47. Electrolytic bath made of plexiglass with an ultrasonic vibrator.

The most satisfactory results described earlier were obtained with these positions of the vibrators.

In one of the plants in the Soviet Union the following baths with capacities of 3-200 liters were developed and tested:

a) A bath with an AVDI vibrator mounted at the bottom;
b) a bath with an AVBI vibrator mounted at the side;
c) a combined bath with AVDBI vibrators at the bottom;
d) a bath with an immersed API vibrator. The bodies of these baths are made of vinyl or other plastic and the vibrator and the membrane are made of nickel.

An experimental industrial apparatus consisting of a UZG-10 generator, a control panel, and two blocks of 200-liter baths with four vibrators was developed on the basis of laboratory results. With these baths the performance per unit working volume of the bath increased by a factor of 4-5.

When powerful ultrasonic generators are used (over 500 kW) general safety rules must be followed.

The generators must be placed in a dry room free from acid or alkaline vapors. The body of the generator must be grounded with a copper tube at least 6 mm in diameter. The cover must be provided with blocking contacts. Work with short-circuited blocking contacts is forbidden. The floor in front of the apparatus must be covered with a rubber mat. The equipment should be dismantled only by persons familiar with the repair and operation of high-frequency equipment.

Some Methods of Producing Electrodeposition of Metals in an Ultrasonic Field

General Recommendations

To select the optimum conditions for producing plating consisting of small crystals which are of uniform thickness, are dense, and have low porosity, one must find the most favorable combination of ultrasonic frequency and intensity for each metal. It was shown experimentally that for metals such as copper, nickel, and zinc a frequency of 18-25 kc and intensities up to $3 \cdot 10^4$ W/m^2 are the most favorable.

The ultrasonic frequency and intensity can be higher for chromium. For electrolytic deposits whose hardness is no higher than that of nickel and chromium deposits (tin, lead, and cadmium, for example) the ultrasonic characteristics must be selected in accordance with the physical properties of these metals.

Ultrasonic vibration improves the quality of electroplating and accelerates the electrodeposition process because higher current densities can be used; this, in turn, makes it possible to use simpler and less concentrated electrolytes, and the acidity of the electrolyte is also stabilized. When an ultrasonic field is used during electrodeposition, the correction of the composition of the electrolyte is simplified and it becomes unnecessary to add different salts to the electrolyte.

The successful application of ultrasonic vibration during electrodeposition is ensured by preliminary treatment of the articles to be plated with ultrasonic vibrations in the electrolytic bath before the current is turned on. This renders the surface cleaner and removes adsorbed films and impurities. The treatment also improves the quality of cathodic deposits.

Ultrasonic vibration is particularly effective when the cathode is oriented parallel to the direction of the ultrasonic field. In all other positions the ultrasonic field is less effective, and this can be utilized when high acoustic intensities are used for electroplating of parts with complicated shapes.

In some cases ultrasonic vibration may have a negative effect on the process of electrodeposition of metals.

TABLE 28. Equipment Used in Industry and in Laboratory Research

Type of equipment	Type of vibrator	Equipment parameters		Some technical characteristics				Uses	Source of data
		Frequency, kc	Intensity, W/m^2	Power consumption, kW	Power output, kW	Dimensions, mm	Weight, kg		
Experimental equipment	Nickel magnetostrictor	21	$5 \cdot 10^4 - 6 \cdot 10^4$	—	—	—	—	Electrodeposition of metal, cleaning, descaling	[11]
Energolegprom plant	Nickel magnetostrictor	23	$1 \cdot 10^4$	2.5	1.5	600×1000	230	Degreasing, electrodeposition of metals	[51]
UZG-10	Magnetostrictor (nickel or permendur)	16-25	$10 \cdot 10^4$	20	9.5	—	620	Applied to liquid systems, electrodeposition of metals	[30]
	Nickel magnetostrictor	30	$1.5 \cdot 10^3 - 3 \cdot 10^3$	—	—	—	—	Electrodeposition of metals	[32]
UZM-1.5	Nickel magnetostrictor	23	$1 \cdot 10^4 - 3 \cdot 10^4$	2.5	1.5	600×1450	275	Transmission of vibrations to the liquid	[63]
BAR	Magnetostrictor (nickel or permendur)	6-100	—	3	0.8	—	—	Electrodeposition of metals	[26]
UZG-3	Magnetostrictor and piezoquartz	12-400	$3 \cdot 10^4$	5	2	645×740	1550	Transmission of vibrations to the liquid	[64]
A624.15	Magnetostrictor	15-30	—	8	8	1190 830 1800	1200	Cleaning of machine parts	[65]

TABLE 28. (Continued)

Type of equipment	Type of vibrator	Equipment parameters		Some technical characteristics				Uses	Source of data
		Frequency, kc	Intensity, W/m^2	Power consumption, kW	Power output, kW	Dimensions, mm	Weight, kg		
High-frequency generator	Piezoquartz	300 300	—	—	1.5	—	—	Dispersion, intensification of physicochemical processes	[64]
High-frequency generator for industrial purposes	Piezoquartz	400 600	—	—	—	—	—	Laboratory investigations	[64]
Laboratory radio-frequency generator	Barium titanate	1400	$1 \cdot 10^4$	—	0.550	—	—	Oxidation-reduction processes	[24]
Generator constructed by MOSKIP	Magnetostrictor, piezoquartz, or barium titanate	25-1000	—	—	—	900 1200 1500	—	—	[66]

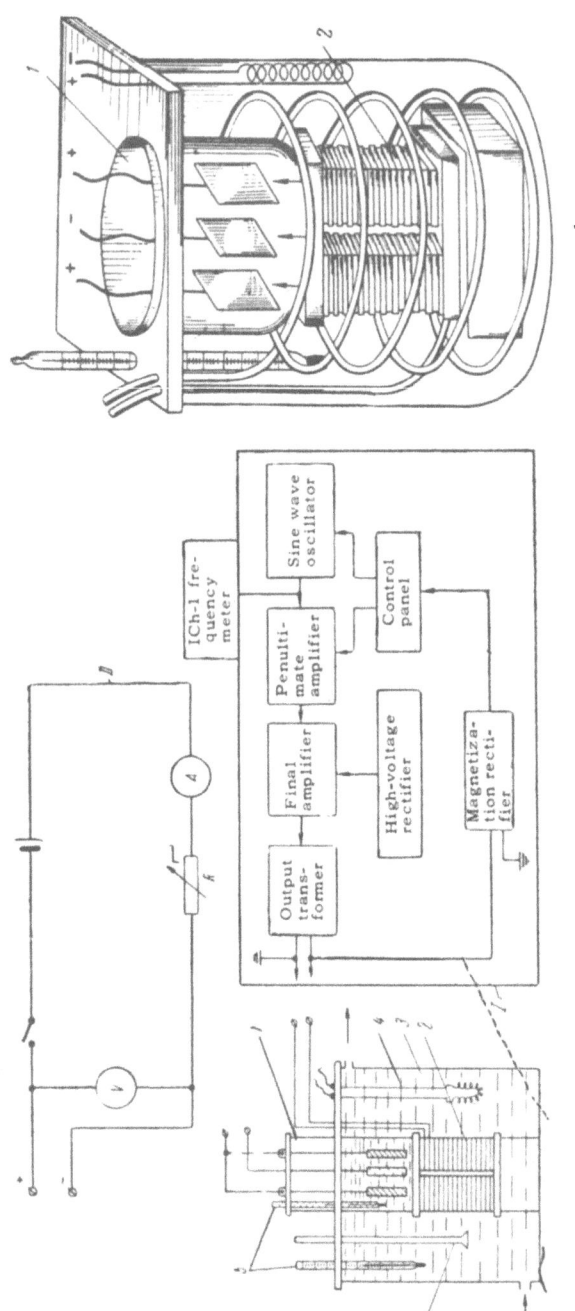

b

Fig. 48. Diagram of experimental equipment for electrodeposition of metals in an ultrasonic field. a) Diagram of the equipment: 1) electrolytic bath; 2), magnetostrictor; 3) thermostat; 4) heater; 5) thermometers; 6) mixer. b) Position of the electrolytic bath and the vibrator in the thermostat.

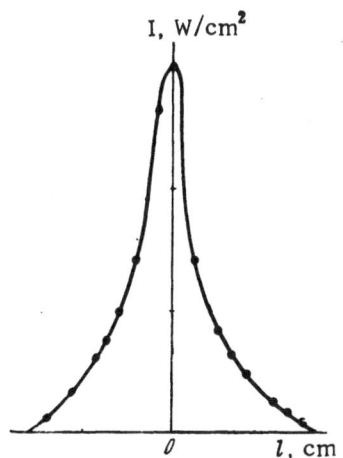

Fig. 49. Distribution of ultrasonic field intensity in the bath.

Electrodeposition of metals is impaired by inadequate ultrasonic characteristics, and then not only the deposit but even the material of the cathode may be broken up. The irregularity of deposition increases and rippled surfaces may be formed. The continuity of the deposits is disrupted and in some cases the deposits are detached from the surface of the cathode.

Incorrect ultrasonic characteristics may lead to porous deposits with large grains, which are caused particularly by intensified cavitation phenomena. Also, the structure of the deposit may become irregular. For example, when copper is deposited in an ultrasonic field with an intensity higher than $1 \cdot 10^5$ W/m^2 the deposit and the cathode itself begin to dissolve. Let us also note that the crumbling of electrolytic deposits under the influence of ultrasonic vibration may be very useful in obtaining powdered metal, which collects at the bottom of the electrolytic bath. Incorrect ultrasonic characteristics affect the anodic process in particular. Anodic dissolution is accompanied by mechanical destruction of the anode, and this increases the expenditure of the anode metal.

The experimental equipment for electrodeposition of metals in an ultrasonic field is somewhat different from the ordinary equipment. Figure 48 shows a diagram of this apparatus and a detailed description of the apparatus is given in the same figure. The apparatus consists of an ultrasonic generator and a frequency meter (I), an apparatus for electrolysis with thermostats, and a magnetostrictor in the bath (II). Cooling columns must be used in the thermostat to avoid the heating of water in the thermostat and, consequently, the electrolyte in the bath, by ultrasonic vibration.

It should be noted that the temperature of the water in the thermostat is a little different from the temperature of the electrolytic bath, and therefore it is necessary to control the temperature of the bath. The thermometer should not be placed near the cathode so as to avoid screening it. When investigations of electrodepositions are made over a wide temperature range it is necessary to have separate thermostats for the bath and the vibrator. It is well known that at high temperatures (100°C) the magnetostrictive properties of the nickel plates of the vibrators are impaired. In this case the electrolytic bath must be a double-wall cylinder connected with the U-8 or TS-20 thermostat.

Experimental results showed that the position of the cathode with respect to the propagating ultrasonic wave and the ultrasonic intensity greatly affect the quality of the deposit. Usually, the cathode is placed in the middle of the electrolytic bath parallel to the ultrasonic vibration, but other arrangements can also be used.

To obtain uniform good-quality plating, the ultrasonic field must be uniformly distributed in the electrolytic bath and the part to be plated should be subjected to a uniform ultrasonic field. Other conditions being equal, this depends on the shape, size, and position of the vibrators with respect to the part to be plated.

The specific ultrasonic intensity, i.e., the intensity per unit volume of electrolyte, is a very important factor.

To select the position of the cathode in the electrolytic bath one must take into account the distribution of the ultrasonic intensity at different points of the wave. The intensity is much higher at the center of the beam than at the edge (Fig. 49). Cavitation effects are particularly strong in the center of the beam. At the edges of the bath cavitation is insignificant and electrodeposition is affected mainly by the sound pressure and by secondary effects (radiation pressure, acoustic streaming). Sonic currents (acoustic streaming) have no time to develop directly at the vibrator but increase at a certain distance from the vibrator. At a relatively great distance from the source of ultrasonic vibration the intensity of the flow decreases again. Acoustic streaming deforms the shape of the wave, and therefore the electrode must be placed at a certain distance from the source of ultrasonic vibration to utilize this effect in electrodeposition.

TABLE 29. Causes of Defects in Deposits Produced in an Ultrasonic Field and Means of Preventing Them

Type of defect	Cause of defect	Method of preventing the defect
Some areas remain unplated; spots on the plated articles	Effect of acoustic shadow, screening	Correct position of the electrodes; do not place foreign objects in the electrolytic bath
Plating has bright and dark spots	Formation of standing waves	Decrease frequency of vibration; select the correct size of electrodes
Irregularities on the surface and "blowholes"; dispersed particles appear in cavitation	Effect of cavitation	Decrease ultrasonic intensity; do not place electrode in the center of the beam
Decrease in the effect of ultrasonic vibration with time	Increase in the temperature of the electrolyte with increasing time of vibration	Cool the magnetostrictors

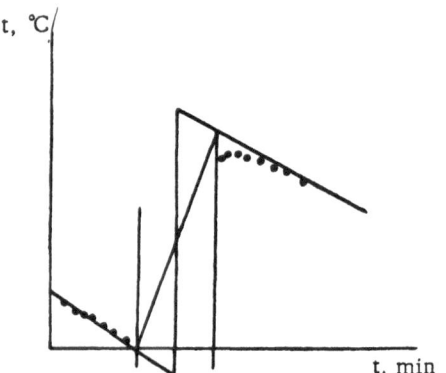

Fig. 50. Determination of ultrasonic field intensity by the calorimetric method.

When using ultrasonic vibration in electrodeposition of metals, one must take into account the singularities of each metal.

There are not yet enough experimental data to give complete recommendations for electrodeposition of a given metal in an ultrasonic field. On the basis of the published data and the results of our experiments we may say that there is an ultrasonic intensity "limit" for each metal, at which the deposits are of good quality. Metals which adhere well to the base metal can be electrodeposited in an ultrasonic field of high intensity. The cathode should be placed in the center of the ultrasonic beam, using the maximum vibrational intensity. Strong cavitation prevents electrocrystallization of metals which adhere poorly to the base metal, and the deposits are detached from the base metal. In this case the cathode should not be placed in the center of the ultrasonic beam. To improve the adhesion of the plating to the base metal the base metal is first cleaned in an ultrasonic field of maximum intensity.

When operating in an ultrasonic field one must take into account such physical properties of the ultrasonic waves as reflection and refraction. Different articles (thermometers, electrolytic switches) placed in the electrolytic bath may create obstacles in the path of the ultrasonic waves.

Ultrasonic waves reflect from and intererfe with the waves in the opposite direction and, as a result, are either intensified or attenuated. An additional irregularity of the field is thus created and is responsible for irregular deposition of the metal. "Blowholes" may appear at the surface of the cathode.

The causes of plating defects and methods of preventing them are given in Table 29.

Thus, correct application of ultrasonic vibration in the electrodeposition of metals gives good results.

Measurements of Intensity

As we noted earlier, the ultrasonic intensity plays an important role in the electrodeposition of metals in an ultrasonic field. There are several methods of measuring the intensity of the ultrasonic field. The simplest is the calorimetric method. In this method one determines the average acoustic energy in space and in time by the amount of heat received by the calorimetric system subjected to ultrasonic vibration.

The calorimeter is a double-walled 200-ml container. Ultrasonic vibration is applied to the flat bottom of the vessel, which must be a single-wall bottom, since air between two walls would absorb the ultrasonic vibrations almost completely. To transform all the ultrasonic energy into heat, the upper part of the calorimeter is covered with rubber which absorbs acoustic vibrations. The temperature of the electrolyte is measured with a Beckmann thermometer. Figure 50 shows the curves obtained from calorimetric measurements. The intensity of the vibration is determined by the formula

$$I = \frac{cm\, \Delta t}{\tau S},$$

where I is the intensity of ultrasonic vibration, W/m^2; c is the specific heat capacity of water, $J/kg \cdot deg$; m is the mass of the water, kg; Δt is the temperature difference, °C; t is the time of application of the ultrasonic field, sec; S is the emitting surface area.

To determine the local intensity of the ultrasonic field one uses thermoacoustic or piezoelectric probes.

LITERATURE CITED

1. P. Morse, Vibration and Sound [Russian translation], GITTL (1949).
2. S. N. Rzhevkin, Lectures on the Theory of Sound, Izd. MGU (1960).
3. V. A. Krasil'nikov, Sonic and Ultrasonic Waves, GITTL (1951).
4. B. B. Kudryavtsev, Application of Ultrasonics in Physicochemical Investigations, GITTL (1952).
5. L. Bergman, Ultrasound [Russian translation], IL (1957).
6. B. B. Kudryavtsev, Ultrasonic Methods of Investigating Substances, Uchpedgiz (1961).
7. O. I. Babikov, Ultrasonic Vibration and Its Applications in Industry, Fizmatgiz (1958).
8. D. A. Gershgal and V. M. Fridman, Ultrasonic Equipment, Gosénergoizdat (1961).
9. B. Carlin, Ultrasonics [Russian translation], IL (1960).
10. A. T. Vagramyan and Z. A. Solov'eva, Methods of Investigating Electrodeposited Metals, Izd. Akad. Nauk SSSR (1960).
11. S. M. Kochergin and N. N. Terpilovskii, "Study of electrocrystallization of metals in an ultrasonic field," Zh. Fiz. Khim. 27:394 (1953).
12. J. Guitton, J. des recherches du centre national de la recherche scientifique 18:199 (1959).
13. A. I. Ryazanov and B. B. Kudryavtsev, Depolarizing Action of Ultrasonic Vibration; Use of Ultrasonic Vibration in Investigations of Substances, No. 13, MOPI (1961).
14. A. I. Ryazanov and B. B. Kudryavtsev, Depolarizing Action of Ultrasound; Use of Ultrasonic Vibration in Investigations of Substances, No. 12, MOPI (1960).
15. E. Yeager and F. Hovorka, "Ultrasonic waves and electrochemistry," J. Acoust. Soc. Am. 25:443 (1953).
16. A. M. Ginberg, Ultrasonic Vibration in Chemical and Electrochemical Processes Used in Machine Construction, Mashgiz (1962).
17. T. Builat, "Ultrasonic vibration in electrodeposition of metals," Metal Finishing 8:65 (1957).
18. G. Schmid and L. Ehret, "Effect of ultrasound on gas liberation potentials," Z. Elektrochem. 43:597 (1937).
19. J. Trillat, "Effect of ultrasonic vibration on anodic oxidation," Compt. Rend. 237:1147 (1953).
20. L. Blum, "Effect of ultrasonic vibration on electrolytic deposition of metals," Rev. Chim. 10:546 (1958).
21. A. Roll, "Effect of ultrasonic vibration on electrode processes," Metal Finishing 9:55 (1957).
22. O. Lindstrom, "Electrochemical effects in a field of stationary ultrasonic waves," Acta Chem. Scand. 6:1313 (1952).
23. L. V. Nikitin, "Sonoelectrochemical phenomena," Dokl. Akad. Nauk 11:63 (1936).
24. F. I. Kukoz and L. A. Kukoz, Use of Ultrasonic Vibration in Investigations of Substances, No. 13, MOPI (1961).
25. A. V. Bondarenko and S. Ya. Popov, "Polarization of cathodes during electrocrystallization of metals in the presence of sonic and ultrasonic vibration," Use of Ultrasonic Vibration in Investigtations of Substances, No. 13, MOPI (1961), p. 87.
26. A. M. Smirnova and N. T. Kudryavtsev, "Investigation of the effect of ultrasonic vibration on electrodeposition of chromium," Zh. Prikl. Khim. 33:2521 (1960).
27. A. M. Smirnova and N. T. Kudryavtsev, "Effect of ultrasonic vibration on the electrodeposition of zinc," Zh. Prikl. Khim. 35:330 (1962).
28. A. M. Ozerov, "Study of the effect of ultrasonic vibration on electrodeposition of chromium," Zh. Prikl. Khim. 35:115 (1962).
29. Yu. M. Bystrov and N. A. Evdokimov, "Effect of ultrasonic vibration on electroplating," Akustika 5:241 (1959).
30. S. A. Shatsova, et al., "Effect of ultrasonic vibration on electrolytic deposition of metals from cyanide baths," Zh. Prikl. Khim. 34:331 (1961).

31. J. Delville, "Effect of ultrasonic vibration on cathodic overvoltage," Compt. Rend. 242:1462 (1956).

32. A. N. Trofimov, "Distribution of metal on the surfaces of cathodes during electrodeposition in an ultrasonic field," Zh. Fiz. Khim. 32:1173 (1958).

33. A. N. Trofimov, Use of Ultrasonic Vibration in Investigations of Substances, No. 12, MOPI (1960).

34. W. Wolfe, H. Chessin, E. Yeager, and F. Hovorka, "Effect of ultrasonic vibrations on the deposition of copper during electrolysis," J. Electrochem. Soc. 101:590 (1954).

35. V. I. Luk'yanov and A. M. Pavlov, "Use of ultrasonic vibration to increase the rate of electrochemical processes in industry," Collection: Application of Ultrasonics in Machine Construction, Izd. TsINTi (1960), p. 80.

36. A. I. Sobolev, "Use of ultrasonic vibration to increase the rate of deposition of metals," Collection: Application of Ultrasonics in Machine Construction, Izd. TsINTI (1960), p. 85.

37. V. A. Mikhailov, New Results in Electrolytic and Ultrasonic Treatment of Metals, Lenizdat (1959), p. 174.

38. F. Müller and H. Kuss, "Effect of a vibrating cathode on the electrolytic deposition of metals," Helv. Chim. Acta 33:217 (1950).

39. P. Mockel, "Chemical effect of Ultrasound," Chem. Czech. 2:71 (1956).

40. S. Hattem, "Effect of ultrasonic vibration on mixtures of aliphatic amino alcohols," Compt. Rend. 229:42 (1943).

41. S. M. Kochergin, et al., "Effect of an ultrasonic field on the penetration of gas in the metal during electrolysis," Nauchn. Dokl. Vysshei Shkola, Khim. i Khim. Tekhnol., No. 4:779 (1958).

42. A. P. Kapustin, "Degassing of liquids in an ultrasonic field," Zh. Tekhn. Fiz. 24:1008 (1954).

43. L. G. Plekhanov, "Effect of ultrasonic vibration on the electrodeposition of metals," Izv. Akad. Nauk Kaz.SSR, No. 1(7):59 (1960).

44. F. Levi, "Effect of ultrasonic vibration on the formation of metal crystals during electrodeposition," Ric. Sci. 19:887 (1949).

45. F. Levi, "Electrolytic deposition of copper in thin layers on a rotating cathode in the presence of ultrasonic waves," Nuovo Cimento 12:493 (1959).

46. V. E. Kavalyunaite, "Effect of ultrasonic vibration on the solution and growth of single crystals," Use of Ultrasonic Vibration in Investigations of Substances, No. 6, MOPI (1958).

47. S. M. Kochergin, et al.,"Use of autoradiographs in the investigation of cathodic deposits of metals," Zh. Fiz. Khim. 32:930 (1958).

48. H. Gollmick and K. Tesser, "Cleaning with ultrasound produced by magnetostrictive vibrators," Metalloberfläche 10:233 (1956).

49. R. Audubert and J. Guitton, "Effect of ultrasonic vibration on the anodic dissolution of metals," Compt. Rend. 242:1458 (1956).

50. A. N. Trofimov, "Effect of ultrasonic vibration on the anodic dissolution of metals," Use of Ultrasonic Vibration in Investigations of Substances, No. 6, MOPI (1958), p. 177.

51. S. M. Kochergin, et al., "Investigation of the anodic dissolution of copper in an ultrasonic field," Zh. Fiz. Khim. 35:917 (1961).

52. A. I. Ryazanov, et al., "Effect of ultrasonic vibration on the anodic dissolution of palladium," Use of Ultrasonic vibration in Investigations of Substances, MOPI (1961), p. 139.

53. G. Schmid and L. Ehret, "Effect of ultrasound on the passivation of metals," Z. Elektrochem. 43:408 (1937).

54. S. Ya. Grilikhes, Preparation of Surfaces of Machine Parts for Electrodeposition, Mashgiz (1961).

55. I. S. Demchuk, Increase of the Rate of Technological Processes by Ultrasonic Vibration, Mashgiz (1960).

56. Collection: Application of Ultrasonics in Machine Construction Technology, Izd. TsINTI (1960).

57. S. M. Kochergin, et al., Tr. Kazansk. Khim. Tekhol. Inst., No. 22:139 (1957).

58. S. Rich, "Improvement in electroplating due to ultrasonics," Plating 42:1407 (1956).

59. W. T. Young and H. Kersten, "An effect of ultrasonic radiation on electrodeposits," J. Chem. Phys. 4:426 (1936).

60. D. Fishock, "Ultrasonic vibration in finishing operations," Metal Ind. 93:109 (1958).

61. T. Rummel and K. Schmitt, "Investigation of the effect of ultrasonic vibration on electrolytic deposition of zinc and copper," Korrosion u. Metallschutz 19:101 (1943).

62. New Results in Electrical and Ultrasonic Treatment of Materials, Lenizdat (1959).

63. Sh. D. Achkinadze, Industrial Application of Ultrasonics in the Construction of Machines and Apparatus, Lenizdat (1958).

64. Use of Ultrasonic Vibration in Investigations of Substances, No. 10, MOPI (1960).

65. V. M. Fridman, Sonic and Ultrasonic Vibrations and Their Use in Light Industry, Gizlegprom (1956).

66. A. M. Ginberg, "Effect of ultrasonic vibration on the electrodeposition of metals," Zh. Vses. Khim. Obshchestva im. D. I. Mendeleeva, 8:502 (1963).

67. V. I. Falicheva et al., "Investigation of the effect of ultrasonic vibration on the deposition of zinc from cyanide electrolytes," Zh. Prikl. Khim. 46:1506 (1963).